GEOMORPHIC PROCESSES I

Anton C. Imeson & Maria Sala (Editors)

GEOMORPHIC PROCESSES
In Environments With Strong Seasonal Contrasts

Vol. I: HILLSLOPE PROCESSES

Selected papers of the "Regional Conference on Mediterranean Countries", IGU Commission on Measurement, Theory and Application in Geomorphology, Barcelona–Valencia–Murcia–Granada, September 5–14, 1986

CATENA SUPPLEMENT 12

CATENA – A cooperating Journal of the International Society of Soil Science

ISSS - AISS - IBG

Cover photo by Claus Dalchow:
Erosion and accumulation on slopes of volcanic sediments. Agricultural landuse is restricted to areas of accumulation, Fuerteventura, Canary Islands, Spain

CIP-Titelaufnahme der Deutschen Bibliothek

Geomorphic processes in environments with strong seasonal contrasts: selected papers of the "Regional Conference on Mediterranean Countries", Barcelona, September 1986. - Cremlingen-Destedt: Catena.
NE: Regional Conference on Mediterranean Countries <1987, Barcelona>
Vol. I. Hillslope processes / Anton C. Imeson; Maria Sala (ed.). - 1988
(Catena: Supplement; 12)
ISBN 3-923381-12-3
NE: Imeson, Anton C. [Hrsg.]; Catena / Supplement

©Copyright 1988 by CATENA VERLAG, D-3302 CREMLINGEN-Destedt, W. GERMANY

All rights are reserved. No part of this publication may be reproduced, stored in a retrieval system or transmitted in any form or by any means, electronic, mechanical, photocopying, recording or otherwise, without prior permission of the publisher.

This publication has been registered with the Copyright Clearance Center, Inc. Consent is given for copying of articles for personal or internal use, for the specific clients. This consent is given on the condition that the copier pay through the Center the per-copy fee for copying beyond that permitted by Sections 107 or 108 of the U.S. Copyright Law. The per-copy fee is stated in the code-line at the bottom of the first page of each article. The appropriate fee, together with a copy of the first page of the article, should be forwarded to the Copyright Clearance Center, Inc., 27 Congress Street, Salem, MA 01970, U.S.A. If no code-line appears, broad consent to copy has not been given and permission to copy must be obtained directly from the publisher. This consent does not extend to other kinds of copying, such as for general distribution, resale, advertising and promotion purposes, or for creating new collective works. Special written permission must be obtained from the publisher for such copying.

Submission of an article for publication implies the transfer of the copyright from the author(s) to the publisher.

ISSN 0722-0723 / ISBN 3-923381-12-3

CONTENTS

Preface

A. Ávila & F. Rodá
Export of Dissolved Elements in an Evergreen-Oak Forested Watershed in the Montseny Mountains (NE Spain) — 1

M. Sala
Slope Runoff and Sediment Production in Two Mediterranean Mountain Environments — 13

J. Sevink
Soil Organic Horizons of Mediterranean Forest Soils in NE-Catalonia (Spain): Their Characteristics and Significance for Hillslope Runoff, and Effects of Management and Fire — 31

A.G. Brown
Soil Development and Geomorphic Processes in a Chaparral Watershed: Rattlesnake Canyon, S. California, USA — 45

T.P. Burt
Seasonality of Subsurface Flow and Nitrate Leaching — 59

K. Rögner
Measurements of Cavernous Weathering at Machtesh Hagadol (Negev, Israel) A Semiquantitative Study — 67

M. Mietton
Mesures Continués des Températures dans le Socle Granitique en Region Soudanienne (Fèvrier 1982–Juin 1983, Ouagadougou, Burkina Faso) — 77

N. La Roca Cervigón & A. Calvo-Cases
Slope Evolution by Mass Movements and Surface Wash (Valls d'Alcoi, Alicante, Spain) — 95

A. Calvo-Cases & N. La Roca Cervigón
Slope Form and Soil Erosion on Calcareous Slopes (Serra Grossa, Valencia) — 103

J. Poesen & D. Torri
The Effect of Cup Size on Splash Detachment and Transport Measurements Part I: Field Measurements — 113

D. Torri & J. Poesen
The Effect of Cup Size on Splash Detachment and Transport Measurements Part II: Theoretical Approach — 127

A.C. Imeson & J.M. Verstraten
Rills on Badland Slopes: A Physico-Chemically Controlled Phenomenon — 139

L.A. Lewis
Measurement and Assessment of Soil Loss in Rwanda — 151

C. Zanchi
Soil Loss and Seasonal Variation of Erodibility in Two Soils with Different Texture in the Mugello Valley in Central Italy — 167

L. Góczán & A. Kertész
Some Results of Soil Erosion Monitoring at a Large-Scale Farming Experimental Station in Hungary — 175

H. Lavee
Geomorphic Factors in Locating Sites for Toxic Waste Disposal — 185

PREFACE

In September 1986, the IGU Commission on Measurement, Theory and Application in Geomorphology (COMTAG) organised its main autumn meeting in Spain. This meeting had as its theme "Geomorphological Processes in environments with strong seasonal contrasts". Most of the papers presented at the meeting are published in two SUPPLEMENTS of CATENA, of which this is the first, grouping papers focussing on hillslope processes. In part two, the companion volume contains papers dealing mainly with catchment and channel studies.

The COMTAG meeting in Spain was organised principally in Barcelona, Murcia and Granada, where the paper sessions were held. A major feature of the meeting was the large number of excursions to sites where field investigations were in progress. In this way the COMTAG symposium drew attention to the interesting and important work being done in Spain and at the same time served to stimulate research in process geomorphology.

The location of the meeting in Spain provided the opportunity for COMTAG to focus its attention on regions having strong seasonal contrasts. Geomorphological systems in such environments are highly complex and difficult to study due to seasonally extreme conditions. The meeting reflected the need to increase our understanding of the basic processes important in these regions; of the effects of seasonality on interactions between abiotic and biotic processes; and on relationships between climatic parameters and processes. A better understanding of the effect of extreme seasonality has application in terms of our appreciation of the effects of drought in less seasonally extreme humid regions.

The sixteen papers which comprise this volume have been arranged according to the emphasis given to measurement, theory, or application. Six papers present results from ongoing research in Spain. The first three from Catalonia refer to work in forested ecosystems where high amounts of winter precipitation make seasonal contrasts relatively high. The paper by AVILA & RODA deals with chemical denudation, in the same area that SALA has measured mechanical erosion. SEVINK considers how processes in these areas are influenced by organic soil horizons. The next four papers by BROWN, BURT, RÖGNER and MIETTON present the results of measurements from studies in respectively England, California, Israel and Burkina Faso. BROWN demonstrates how studies of magnetic susceptibility can increase our insight into processes in Mediterranean environments; RÖGNER describes measurements of Cavernous weathering in the Negev; BURT draws attention to the seasonality of nitrate leaching in the relatively less seasonal environment of southwest England; and MIETTON addresses the technical problems of measuring temperature in the hot dry environment of Burkina Faso. The last two papers, dealing mainly with method both by LA ROCA & CALVO CASES, describe measurements from marls in Valencia. The role of extreme events in this region in influencing process and morphological development is one of the points stressed.

In two papers POESEN & TORRI describe how cup size influences splash transport and detachment. This is followed by IMESON & VERSTRATEN who consider how rill initiation is related to physico-chemically controlled soil properties that influence the dynamic response of material to wetting.

Three of the final four papers concerned with application deal with soil erosion based on different approaches. LEWIS, working in Rwanda using simple measurements and the USLE, looks mainly at the effect of groundcover on soil loss; ZANCHI in Central Italy uses the USLE with plot measurements of soil loss to draw attention to the importance of seasonal variations in soil erodibility. Plot measurements also form the basis of the paper by GOCZAN & KERTESZ who present the result of measurements from Hungary. Finally LAVEE shows how geomorphic principles can be applied to locate sites for toxic waste disposal.

Two papers presented during the symposium are not included because of space but will be published elsewhere. These are by HAIGH on the Environmental correlates of landslide frequency along new highways in the Central Himalaya and by GERITS on the implications of chemical thresholds and physical-chemical processes for modelling erosion in Spain.

A.C. Imeson
M. Sala

EXPORT OF DISSOLVED ELEMENTS IN AN EVERGREEN-OAK FORESTED WATERSHED IN THE MONTSENY MOUNTAINS (NE SPAIN)

A. Ávila & F. Rodà, Bellaterra (Barcelona)

Summary

Dissolved element inputs in bulk precipitation and outputs in streamwater have been measured for one year in a small, evergreen-oak forested watershed on metamorphic schists at La Castanya (Montseny). This watershed is a sink for atmospheric H^+, NH_4^+ and NO_3^-. Inputs of inorganic nitrogen in bulk precipitation plus dry fallout were 5.7 kg/ha/yr, while outputs in streamwater were only 0.1 kg/ha/yr. This watershed is a source of Na^+ and Mg^{2+}, with net outputs of 17.8 and 5.7 kg/ha/yr, respectively. These net outputs must come from the chemical weathering of silicates, and there is a concurrent large export of HCO_3^-. A net output of 4.3 kg SO_4^{2-} -S/ha/yr could be derived from dry deposition of S or from the oxidation of metallic sulfides in the rock. Potassium, Ca^{2+} and Cl^- appear to be in steady-state at the watershed level. From the annual net output of Na^+ and the Na^+ content of these schists, a rock weathering rate of 2350 kg of rock/ha/yr is estimated. From the net output of cations, a dissolved denudation rate of 1.3 keq/ha/yr is obtained for this watershed, a figure 60% higher than the average rate of a literature compilation of boreal and temperate forested watersheds.

Resumen

Se han medido durante un año las entradas (agua de lluvia) y las salidas (agua del torrente) de elementos disueltos en una pequeña cuenca de La Castanya (Montseny) tallada en esquistos y cubierta por un bosque de encinas. Esta cuenca es un sumidero de H^+, NH_4^+ y NO_3^- atmosféricos. Las entradas de nitrógeno inorgánico en la precipitación fueron de 5.7 kg/ha/año, mientras que las salidas en el agua del torrente fueron sólo de 0.1 kg/ha/año. Esta cuenca es una fuente de Na^+ y Mg^{2+}, con salidas netas de 17.8 y 5.7 kg/ha/año respectivamente. Estas salidas deben de proceder de la meteorización química de los silicatos; existe también una salida importante de HCO_3^-. La salida de 4.3 kg de SO_4^{2-} -S/ha/año puede proceder de la deposición seca de S o de la oxidación de sulfuros metálicos de la roca. El potasio, Ca^{2+} y Cl^- aparecen en equilibrio a nivel de la cuenca. Por la salida anual de Na^+ y por el contenido en Na^+ de los esquistos, se estima que la tasa de meteorización de la roca es de 2350 kg/ha/año. De la salida neta de cationes se obtiene una tasa de denudación por

ISSN 0722-0723
ISBN 3-923381-12-3
©1988 by CATENA VERLAG,
D-3302 Cremlingen-Destedt, W. Germany
3-923381-12-3/88/5011851/US$ 2.00 + 0.25

disolución de 1.3 keq/ha/año, un valor 60% más alto que la tasa media obtenida en cuencas forestales de climas boreales y templados.

1 Introduction

Elemental outputs in water draining a watershed result from atmospheric inputs and from biogeochemical processes within the watershed. Undisturbed forest ecosystems control to a large degree the amount and pathways of water flowing through the watershed, and also the amounts and form of exported materials (e.g. BORMANN & LIKENS 1969). Through such biotic regulation, losses of elements crucial to ecosystem function are often minimised (GORHAM et al. 1979). Besides, in undisturbed forest watersheds where subsurface flow is usually the main contributor to runoff, dissolved matter is often the major form of elemental export in drainage water (e.g. LIKENS et al. 1977). Hence, in humid and subhumid climates, output of dissolved elements in streamwater is frequently the dominant current process of denudation in forested watersheds.

Elemental budgets at the watershed level have been scarcely studied in Mediterranean forest ecosystems. To our knowledge, there is hitherto only one such study, that of ESCARRÉ et al. (1984a, b) in an evergreen-oak (*Quercus ilex*) watershed in the Prades mountains (Tarragona, NE Spain). Mediteranean-type climates are strongly seasonal, with prolonged summer drought and often erratic rainfall patterns. Mediterranean forests are usually water-limited. Given the close relationship between water flow and dissolved export it seems interesting to ask wether this seasonality and low water-availability are reflected or not in the rates of chemical denudation in Mediterranean forest ecosystems, as compared with temperate forests.

We report here the gross and net export rates of dissolved elements for the first year of study in a small, evergreen-oak forested watershed in the Montseny mountains.

2 The Study Area

The experimental watershed (TM9) is located in La Castanya (Montseny mountains, Barcelona; 41° 46' N, 2° 21'E). It is a north-facing, 4.3 ha waterhsed, ranging in elevation from 700 to 1035 m a.s.l., and with a mean slope of 36°. The bedrock is a metamorphic schist formed mainly by quartz, sericite, albite and chlorite. Soils are colluvial rankers, 0.4–1.5 m deep. Mean annual precipitation for the watershed is estimated to be around 900 mm (RODÁ 1983). Precipitation measured during 4 years at the watershed outlet averaged 911 mm/year. Mean annual air temperature is estimated to be around 9°C. The watershed is totally covered by a dense forest of evergreen-oak. The forest was heavily coppiced for charcoal production until 25–30 years ago, and it has not been appreciably disturbed since then. The forest is currently accreting biomass. At the watershed outlet, the stream is permanent.

3 Methods

Streamflow is continuously recorded at a gaging station at the watershed outlet. The weir is a thin-crest, 60° V-notch. The amount of precipitation is measured weekly with a standard rain gage beside the weir. Rainwater is collected for chemical analyses in 4 polyethylene, funnel-type collectors continuously open to the atmosphere. A bucket, lined with a clean polyethylene bag, is used to determine the amount and chemistry of occasional snowfalls. A streamwater sampler, automatically switched on when the water reaches a preselected level, takes high-frequency samples during storm events.

The instrument station is visited at least once a week to collect grab samples from the stream and to change the rain collectors for new ones, thoroughly cleaned in the laboratory with diluted Cl and distilled water. Soluble material deposited as dust onto the collector funnels is recovered in separate bottles by brushing the inner side of the funnel with a clean nylon mesh and rinsing the funnel with a small amount of distilled water.

Upon arrival to the laboratory, conductivity and pH are measured electrometrically, and alkalinity by conductimetric titration. The other analyses are usually performed on samples stored at $-20°C$. Sodium and K^+ are analyzed by flame emission, and Ca^{2+} and Mg^{2+} by atomic absorption spectrometry, after addition of 400 ppm La and 1% HCl to blanks, standards and samples. Readings of the four cations are made at the Servei d'Espectroscòpia de la Universitat de Barcelona. Ammonium, NO_3^-, SO_4^{2-} and Cl^- are analyzed by ion chromatography.

The annual input of each element in bulk deposition was calculated by multiplying its volume-weighted annual mean concentration in precipitation by the total measured precipitation. The amounts of elements recovered by washing the funnels were calculated from the concentrations and rinsing volumes. Hydrogen ion concentrations were derived from pH values.

The annual output of each element was calculated as the sum of the outputs for each sample. Element output in a given sample was calculated by multiplying its concentration in the sample by the amount of runoff that occurred during the period from the time half-way to the preceeding sample to the time half-way to the next sample. For this output estimate to be accurate is necessary that storm events be sampled intensively enough to account for streamwater chemistry changes at high flows.

Results reported here refer to the 1-year period beginning in 1 August 1984.

4 Results and Discussion

4.1 Water budget

During the study year, precipitation at TM9 amounted to 870 mm, a figure close to the estimated average rainfall for this watershed. Runoff amounted to 404 mm. An evergreen-oak forested watershed in Prades, with roughly similar bedrock and soils, produced in 1 year of study only 82 mm of runoff with a precipitation of 577 mm (ESCARRÉ et al. 1984a.) Interestingly, the evapotranspiration estimates (obtained by difference between annual precipitation and runoff) are very similar at Montseny and Prades: 466 and 495 mm, respectively. Interception losses are included in these estimates.

Monthly variations of precipitation and runoff at TM9 are shown in fig.1. November and May were the wettest months in this study, accounting between them for 48% of the annual precipitation and 52% of the annual runoff.

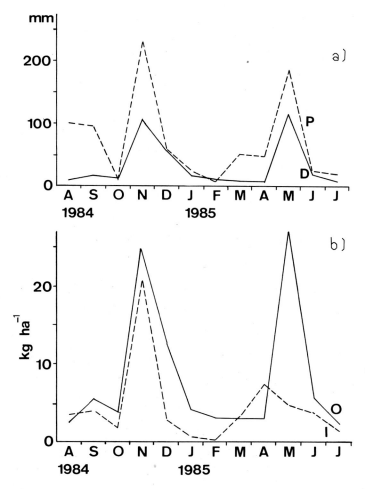

Fig. 1: *a: Monthly budget of water inputs in precipitation (P) and water outputs in runoff (D) in an evergreen-oak forested watershed. b: Monthly budget of inputs in bulk deposition (I) and outputs in streamwater (O) of total dissolved elements analyzed in this study.*

4.2 Monthly Outputs of Dissolved Elements

Monthly outputs of total dissolved elements at TM9 have a closer relationship with monthly outputs of water than with monthly inputs of dissolved material (fig.1). Two factors may account for this fact. First, elemental output is the product of runoff by element concentration in streamwater; since runoff is much more variable than ionic concentrations, output depends mainly on amount of runoff. Second, elemental inputs are processed within the system, outputs being partly the result of biogeochemical control mechanisms in the vegetation and in the soil. For example, precipitation can be abundant but very diluted, as in May 1985 at TM9 (fig.1), giving low elemental inputs but high runoff and, therefore, high output of dissolved elements. Due to this dependence of gross export on runoff, November and May account for 53% of the total dissolved output during the study year at TM9. Hence, most of the material exported from Mediterranean watersheds in streamwater should occur during the rainiest months, somewhere between October and May. Large interannual variability in elemental budgets is expected in these watersheds because of the erratic rainfall pattern.

The above mentioned positive correlation between monthly totals of runoff and element export at TM9 holds for all the analyzed ions ($r^2 > 0.9$, $P < 0.01$ in all cases). As often occurs, the correlation is lowest for NO_3^- ($r^2 = 0.91$). Flow-weighted monthly mean NO_3^- concentrations in streamwater at TM9 were below or just above the analytical detection limit (2 μeq/L) for every month except July and August, when they reached 3.9 and 3.5 μeq/L, respectively. Hence, for a given amount of monthly runoff, NO_3^- outputs are higher in July and August than in the rest of the year. This result is opposed to the common situation in many temperate forests where low streamwater NO_3^- concentrations during the growing season are attributed to biological uptake of nitrogen (e.g. FELLER 1981, LIKENS et al. 1977, REHFUESS 1981). At TM9, an examination of the NO_3^- concentrations in individual samples throughout the year reveals that NO_3^- was almost always below the detection limit at baseflow, and often so at highflow. Some storm events gave detectable NO_3^-, at least during part of the event. Maximum NO_3^- values (up to 180 μeq/L) were found during short events resulting from intense summer thunderstorms, a fact that is reflected in the weighted means of July and August. Therefore, we think that the higher mean NO_3^- concentrations in these months are **not** due to decreased biological uptake of nitrogen by the evergreen-oak trees during the summer drought period but, instead, that they result from the hydrological properties of these short events, as discussed by ÀVILA & RODÀ (1985).

4.3 Annual Input-output Budgets

Input and output fluxes for the analyzed elements during the study year at TM9 are presented in tab.1. Atmospheric inputs in this table are the result of adding to bulk deposition fluxes the amounts of elements recovered by washing the funnels. These amounts were less than 10% of the bulk deposition input for all the analyzed elements. Biological N_2 fixation, aerosol impaction onto the canopies, and gaseous absorption are not included in the input fluxes of tab.1.

On a mass basis, gross elemental output in streamwater at TM9 is dominated by Na^+, followed by Ca^{2+}, HCO_3^--C, Cl^- and SO_4^{2-}-S, in this order (tab.1). Gross output of NH_4^+-N and NO_3^--N is very low.

Net output from the watershed is gross output minus atmospheric input. Net outputs at TM9 are positive (output larger than input) for all the analyzed ions, except H^+, NH_4^+ and NO_3^- (tab.1). This watershed is a net sink for the last three ions, unless there are important unmeasured outputs. Concerning H^+, the free acidity of incoming precipitation has been completely neutralized when streamwater leaves TM9. During the study year, mean volume-weighted pH in bulk precipitation was 4.73, while

Flux	H^+	Na^+	K^+	Ca^{2+}	Mg^{2+}	NH_4^+-N	NO_3^--N	SO_4^{2-}-S	Cl^-	HCO_3^--C
Input[1]	0.16	6.7	1.4	19.0	1.8	2.9	2.8	8.6	13.1	-[2]
Output	0.00	24.5	1.6	19.2	7.5	0	0.1	12.9	14.3	16.6
Net output	-0.16	17.8	0.2	0.2	5.7	-2.9	-2.7	4.3	1.7	-

[1] Input is bulk precipitation plus dry fallout.
[2] Not measured. Probably less than 5 kg/ha/yr.

Tab. 1: *Input in bulk deposition and output of dissolved elements in streamwater in a small, evergreen-oak forsted watershed in the Montseny mountains. Net output is output minus input. In kg/ha/yr.*

	H^+	Na^+	K^+	Ca^{2+}	Mg^{2+}	NH_4^+-N	NO_3^--N	SO_4^{2-}-S	Cl^-
Evergreen-oak watersheds									
La Castanya	-0.16	17.8	0.2	0.2	5.7	-2.9	-2.7	4.3	1.7
Prades	-0.07	5.6	0.7	22.6	4.9	-3.2[1]		2.5	2.8
Boreal and temperate watersheds									
$\bar{x}$	-0.42	5.73	1.18	6.45	2.91	-2.04	-1.31	0.09	1.79[2]
s_x	0.26	5.47	2.07	9.36	3.16	1.60	2.23	3.85	5.02
n	7	31	31	31	31	26	27	18	14

[1] NH_4 - N + NO_3 - N
[2] Excluding a watershed near the sea (BIRKENES, CHRISTOPHERSEN et al. 1982) with high Cl net output, presumably due to sea aerosol impaction, the average decreases to 0.53 ± 1.78.
[3] Data from: BERGSTRÖM & GUSTAFSON 1985, CALLES 1983, CHRISTOPHERSEN et al. 1982, COSBY et al. 1985, ERIKSSON 1974, FELLER 1981, FELLER & KIMMINS 1984, GOSZ 1980, JOHNSON & SWANK 1973, LEWIS & GRANT 1979, LIKENS et al. 1977, MILLER 1963, PACES 1985, ROSÉN 1982, SCHINDLER et al. 1976, SOLLINS et al. 1980, VERSTRATEN 1977.

Tab. 2: *Net export of dissolved elements from two evergreen-oak forested watersheds, compared with literature averages from boreal and temperate forested watersheds on silicate bedrocks. In kg/ha/yr.*

in streamwater it was 7.45. Many processes may consume protons within a watershed. At TM9, we think that the weathering of silicates and the exchange of cations on the surface of soil colloids are probably the major proton consuming processes. Concerning nitrogen, the accreting evergreen-oak forest at TM9 efficiently removes inorganic N in bulk precipitation, a common finding in the literature (tab.2).

Net output from TM9 is dominated by Na^+ and probably by HCO_3^--C, followed by Mg^{2+} and SO_4^{2-}-S (tab.1). These four elements are the only ones among those analyzed with outputs substantially larger than inputs. Their net exports must come from a source within the watershed or from unmeasured inputs. The source of Na^+ and Mg^{2+} at TM9 is presumably the hydrolisis of silicates (albite and chlorite, respectively), a process that consumes protons, releases basic cations to the soil solution and, if H_2CO_3 from soil respiration is acting as the proton donor, produces HCO_3^-.

Net export of S at TM9 probably comes either from the oxidation of metalic sulfides present in the bedrock or from dry deposition of S, or both.

Potassium, Ca^{2+} and Cl^- show only limited net outputs from TM9 (tab.1). Having in mind the large interannual variability often reported in watershed budgets (e.g. LIKENS et al. 1977) and the methodological uncertainties involved, we think that the low net export of these three elements is not enough to reject the hypothesis that they are in steady-state at TM9. For Cl^-, this was an expected result because there are not major sinks or sources of this element within the system. As revealed by throughfall studies in this forest, marine aerosol impaction is of minor importance (FERRÉS et al. 1984, RODÀ 1983) and contributes little Cl^- to the system. Thus, input and output fluxes of Cl^- seem to be in reasonable balance at TM9. From this finding we infer that water entering or leaving the watershed through subsurface flow across the watershed boundaries and water leaving it by deep sepage is negligible in the hydrological and biogeochemical budgets of TM9, a prerequisite in budgetary studies in experimental watersheds.

The almost nil net export of Ca^{2+} from TM9 is somewhat surprising since in many other forested watersheds on silicate bedrock, net Ca^{2+} output is similar to or greater than net output of Na^+ (tab.2). This low net output of Ca^{2+} may perhaps be an anomalous situation for the particular year here reported, because bulk deposition input of Ca^{2+} during the study year was 17.7 kg/ha/yr, almost doubling the average Ca^{2+} input of three previous years at TM9 (9.6 kg/ha/yr; ÀVILA 1986, FERRÉS et al. 1984). This difference arises largely because during the study year two rain events of 50 and 31 mm, very rich in Ca^{2+}, were responsible for 40% of the annual Ca^{2+} input in bulk deposition. If annual Ca^{2+} output in streamwater were relatively independent of Ca^{2+} input in the same year, then a large Ca^{2+} input during the study year would give a lesser net Ca^{2+} output than the long term average. However, the low figure for net Ca^{2+} export could also result from a low rate of Ca^{2+} release from silicate weathering, as discussed below.

4.4 Weathering Rates

The amounts of metallic elements released by weathering of primary silicates at TM9 were estimated from the Na^+ budget. This assumes that the annual release of Na^+ by weathering equals the net export of Na^+ from the watershed plus the small amount of Na^+ annually accumulated in the accreting biomass. As a first approach, it is also assumed that all the involved cation-containing minerals weather at the same rate than the Na-bearing minerals. Other cations could be used as a reference, but Na^+ has the advantage of being incorporated in secondary minerals or accumulated in the plant biomass to a much lesser degree than K^+, Ca^{2+} or Mg^{2+}. Weathering rates at the watershed level have been estimated from Na^+ budgets by ESCARRÉ et al. (1984b), FELLER (1981), JOHNSON et al. (1986) and PACES (1985), among others.

Annual net Na^+ export from TM9 is 17.8 kg/ha (tab.1). As stated above for Cl^-, the impaction of sea-salt aerosols onto the canopies of this forest is only of minor importance, and so it makes little difference wether the annual net Na^+ export from TM9 is corrected or not for aerosol impaction. The net annual increment of Na^+ stored in the above-ground biomass of an evergreen-oak stand adjacent to TM9 was found to be 0.06 kg/ha (FERRÉS 1984). Below-ground Na^+ increment is not available but should be even less than the latter figure. Hence, annual release of Na^+ from weathering at TM9 is taken to be 17.9 kg/ha. Metallic cation contents of the bedrock were computed from CASAS (1979), who analyzed 400 samples of Montseny phyllites and schists. From these, we selected his results of 4 rock samples from La Castanya. The Ca^{2+} value for one of these samples was discarded because it seemed to us anomalously low. Average cation contents of the 3–4 samples used were, in percent of rock mass: Na^+ 0.76, Ca^{2+} 0.15, K^+ 2.98 and Mg^{2+} 0.60. With these figures, 2350 kg/ha of rock must weather annually to provide those 17.9 kg/ha of Na^+.

According to GREGOR (1970), estimates of rock weathering on a worldwide basis vary from 660 to 810 kg/ha/yr. FELLER (1981), using a method like ours, estimated the dacite bedrock of two *Eucalyptus*-forested watersheds in Victoria (Australia) to weather at a rate of 490–860 kg/ha/yr. FELLER (1981) also estimated with data from GUTHRIE et al. (1978) the rock weathering rate of another *Eucalyptus* forested watershed on Ordovician sedimentary bedrock to lie in the range of 500-900 kg/ha/yr.

ESCARRÈ et al. (1984b) and JOHNSON et al. (1968) estimated rock weathering rates by the method of BARTH (1961) that takes into account the Na concentration in the soil besides the abundance in the bedrock and the net Na output, and hence their figures are not strictly comparable to ours.

The complete weathering of 2350 kg/ha/yr of rock at TM9 should release 3.5 kg of Ca^{2+}, 70 kg of K^+ and 14 kg of Mg^{2+} per ha and year. Thus, only a limited amount of Ca^{2+} is estimated to be relased from weathering at TM9. As we have seen, input and output for this element are closely balanced for the study year (tab.1). However, Ca^{2+} is being accumulated in the biomass of this forest. The exact rate of accumulation in the watershed vegetation is not known but must be less than 23 kg/ha/yr, the estimated accumulation rate for a nearby stand (FERRÉs 1984) lying at the base of the slope, with deeper soils and higher water availability, and presumably, higher productivity. Even if Ca^{2+} biomass accumulation is less than 23 kg/ha/yr, it seems to us that the estimated Ca release from rock weathering is not enough to account for the forest requirements. Where does the unaccounted Ca^{2+} come from? Throughfall studies indicate that aerosol impaction could provide at most 4 kg/ha/yr in this forest (FERRÉS et al. 1984). Furthermore, the true rate of weathering release of Ca^{2+} at TM9 may have been underestimated by our method. First, calcium-rich silicates could weather at higher rates than albite does. Second, part of the Ca^{2+} of these schists could be contained in carbonates (CASAS 1979), being thus solubilized faster than Na^+. Another possibility is that trees obtain their needs of Ca^{2+} from exchangeable Ca^{2-} in the soil. This would result in a decrease of Ca^{2+} in the exchange sites. However, the size of the exchangeable Ca^{2+} pool in this forest lies in the range of 1000–20000 kg/ha (A. HERETER, personal communication). This amount seems large enough to provide the Ca^{2+} increment in the plant biomass throughout the aggrading phase of the forest without problems of plant nutrition or soil acidification.

For Mg^{2+} and, particularly, for K^+, the estimated weathering release at TM9 is large compared to the sum of their net output (tab.1) and their annual increment in the biomass, which must be somewhere below 3.3 and 14.7 kg/ha respectively, the accumulation rates for the evergreen-oak stand at the base of the slope (FERRÉS 1984). Probably, chlorite and sericite (the major Mg- and K-bearing minerals in these schists, respectively) weather at a lower rate than albite. Furthermore, weathered K^+ may be incorporated in secondary minerals, as illite-type clays, being thus not exported from the watershed.

4.5 Cationic Denudation Rates

Estimates of rock weathering rates are subjected to considerable methodological uncertainties. More straightforward is the use of dissolved denudation rates defined as the sum of the net outputs of cation equivalents. Such estimates are possible in experimental watersheds where detailed ionic budgets are available. Water pollution sources should not occur within the watershed if the derived denudation rate is to be biogeochemically meaningful.

At TM9, the cationic denudation rate, computed from the net export of Na^+, K^+, Ca^{2+} and Mg^{2+}, is 1.3 keq/ha/yr. With the same criteria, we have calculated an average cationic denudation rate of 0.82 keq/ha/yr (S.D. 0.89 keq/ha/yr) from published data of 31 boreal and temperate forested watersheds on silicate bedrocks (references in tab.2). In this compilation, where evergreen-oak forested watersheds have been not included for comparison purposes, denudation rates for individual watersheds range from -0.1 to 4.4 keq/ha/yr. The estimated cationic denudation rate at TM9 lies within this range, but it is 60% higher than the compilation average. The cationic denudation rate in the evergreen-oak forested watershed at Prades is 1.8 keq/ha/yr (ESCARRÉ et al. 1984a), higher than at Montseny and double than the compilation average.

Although there is at present only one year of data available at both Montseny and Prades, the above figures suggest that cationic denudation rates in these Mediterranean forested watersheds are of the same order of magnitude or higher than denudation rates prevailing in boreal and temperate forest ecosystems. One could have guessed that the recurrence of long periods of low water-availability under a Mediterranean climate would have resulted in low cationic denudation rates through water limitation of either chemical weathering reactions or transport of weathered products. This is clearly not supported by the data from Montseny and Prades. Conversely, some factors may favour increased denudation rates in Mediteranean watersheds, namely, the relatively high annual mean temperature and, perhaps, the alternance of drying/rewetting cycles in the soil.

It is really surprising that the watershed at Prades, which has only one-fifth of the annual runoff of the Montseny watershed, presents a clearly higher cationic denudation rate than the latter. Though climate is drier and warmer at Prades, and average soil thickness is probably higher in Prades, the reasons behind this biogeochemical difference must await further study.

Acknowledgements

This work has been funded by the Comisión Asesora para la Investigación Científica y Tecnica (project 2129/83) and by the Comissió Interdepartamental de Recerca i Innovació Tecnològica. The aid of Teresa Arasa was invaluable. Ana Ma. Gascó helped with the chemical analysis.

References

ÀVILA, A. (1986): Composición química de la lluvia en el Montseny. Actas de las Jornadas sobre las Bases Ecológicas de la Gestión del Medio Terrestre. Diputació de Barcelona. Barcelona.

ÀVILA, A. & RODÀ, F. (1985): Variaciones del quimismo del arroyo durante las crecidas en una cuenca de encinar montano. Cuadernos de Investigación Geográfica 11, 21–31.

BARTH, T.F.W. (1961): Abundance of elements, areal averages and geochemical cycles. Geochimica and Cosmochimica Acta 23, 1–8.

BERGSTRÖM, L. & GUSTAFSON, A. (1985): Hydrogen budgets of four small runoff basins in Sweden. Ambio 44, 346–348.

BORMANN, F.H., LIKENS, G.E. & EATON, J.S. (1969): Biotic regulation of particulate and solution losses from a forested ecosystem. BioScience 19, 600–610.

CALLES, U.M. (1983): Dissolved inorganic substances. A study of mass balance in three small drainage basins. Hydrobiologia 101, 13–18.

CASAS, A. (1979): Estudio litogeoquímico del Paleozoico del Montseny. Doctoral dissertation. Universitat de Barcelona.

CHRISTOPHERSEN, N., SEIP, H.M. & WRIGHT, R.F. (1982): A model for streamwater chemistry at Birkenes, Norway. Water Resources Research 18, 977–996.

COSEY, B.J., WRIGHT, R.F., HORNBERGER, G.M. & GALLOWAY, J.M. (1985): Modeling the effects of acid deposition: estimation of long-term water quality responses in a small forested catchment. Water Resources Research 21, 1591–1601.

ERIKSSON, E. (1974): Vattnet-Kemikaliebäraren. Forsking och Framstag 5, 41–45.

ESCARRÈ, A., GRACIA, C., RODÀ, F. & TERRADAS (1984a): Ecologia del bosque esclerófilo mediterráneo. Investigación y Ciencia 95, 68–78.

ESCARRÈ, A., LLEDÓ, M.J., BELLOT, J., SANCHEZ, J.R., ESCLAPÉS, A., CLEMENTE, A. & ROVIRA, A. (1984b): Compartimentos y flujos biogeoquímicos en un encinar. Distinto predominio de factores físicos y biológicos en su control. Comunicación a la IV Conferencia Internacional de Ciencias Naturales. La Habana.

FELLER, M.C. (1981): Catchment nutrient budgets and geological weathering in Eucalyptus regnans ecosystems in Victoria. Australian Journal of Ecology 6, 65–77.

FELLER, M.C. & KIMMINS, J.P. (1984): Effects of clearcutting on streamwater chemistry and watershed nutrient budgets in southern British Columbia. Water Resources Research 20, 29–40.

FERRÉS, L. (1984): Biomasa, producción, y mineralomasas del encinar montano de La Castanya (Montseny, Barcelona). Doctoral dissertation. Universitat Autònoma de Barcelona.

FERRÉS, L., RODÀ, F., VERDÚ, A.M.C. & TERRADAS, J. (1984): Circulación de nutrientes en algunos ecosistemas forestales del Montseny. Mediterranea, Ser. Biologia 7, 139–166.

GORHAM, E., VITOUSEK, P.M. & REINERS, W.A. (1979): The regulation of chemical budgets over the course of terrestrial ecosystem successions. Annual Review of Ecology and Systematics 10, 35–84.

GOSZ, J.R. (1980): Nutrient budget study for forests along an elevational gradient in New Mexico. Ecology 61, 515–521.

GREGOR, B. (1970): Denudation of the continents. Nature 228, 273–275.

GUTHRIE, H.B., ATTIWILL, P.M. & LEUNING, R. (1978): Nutrient cycling in a Eucalyptus obliqua (L'Herit) forest. II. A study in a small catchment. Australian Journal of Botany 26, 189–201.

JOHNSON, N.M., LIKENS, G.E., BORMANN, F.H. & PIERCE, R.S. (1968): Rate of chemical weathering of silicate minerals in New Hampshire. Geochimica and Cosmochimica Acta 32, 531–545.

JOHNSON, P.L. & SWANK, W.T. (1973): Studies of cation budgets in the southern Appalachians on four experimental watersheds with contrasting vegetation. Ecology 54, 70–80.

LEWIS, W.M. & GRANT, M.C. (1979): Changes in the output of ions from a watershed as a result of the acidification of precipitation. Ecology **60**, 1093–1097.

LIKENS, G.E., BORMANN, F.H., PIERCE, R.S., EATON, J.S. & JOHNSON, N.M. (1977): Biogeochemistry of a forested ecosystem. Springer-Verlag, New York.

MILLER, R.B. (1963): Plant nutrients in hard-beech. The immobilization of nutrients. New Zealand Journal of Science **6**, 365–377.

PACES, T. (1985): Sources of acidification in Central Europe estimated from elemental budgets in small basins. Nature **315**, 31–36.

REHFUESS, K.E. (1981): Über die Wirkungen der Sauren Niederschläge in Waldökosystemen. Forstwissenschaftliches Centralblatt **6**, 363–381.

RODÀ, F. (1983): Biogeoquímica de les aigües de pluja i de drenatge en tres ecosistemes forestals del Montseny. Doctoral dissertation. Universitat Autònoma de Barcelona.

ROSÉN, K. (1982): Supply, loss and distribution of nutrients in three coniferous watersheds in Central Sweden. Reports in Forest Ecology and Forest Soils. 41. Department of Forest Soils. Swedish University of Agricultural Sciences. Uppsala.

SCHINDLER, D.W., NEWBURY, R.W., BEATY, K.G. & CAMPBELL, P. (1976): Natural water and chemical budgets for a small precambrian lake basin in Central Canada. Journal of Fisheries Research Board of Canada **33**, 2526–2543.

SOLLINS, P., GRIER, G.C., McCORISON, F.M., CROMACK, K. & FOGEL, R. (1980): The internal element cycles of an old-growth douglas-fir ecosystem in western Oregon. Ecological Monographs **50**, 261–285.

VERSTRATEN, J.M. (1977): Chemical erosion in a forested watershed in the Oesling, Luxembourg. Earth Surface Processes **2**, 175–184.

Address of authors:
Anna Àvila and Ferran Rodà
Laboratori d'Ecologia
Universitat Autònoma de Barcelona
08193 Bellaterra (Barcelona), Spain

H.-R. BORK u. W. RICKEN

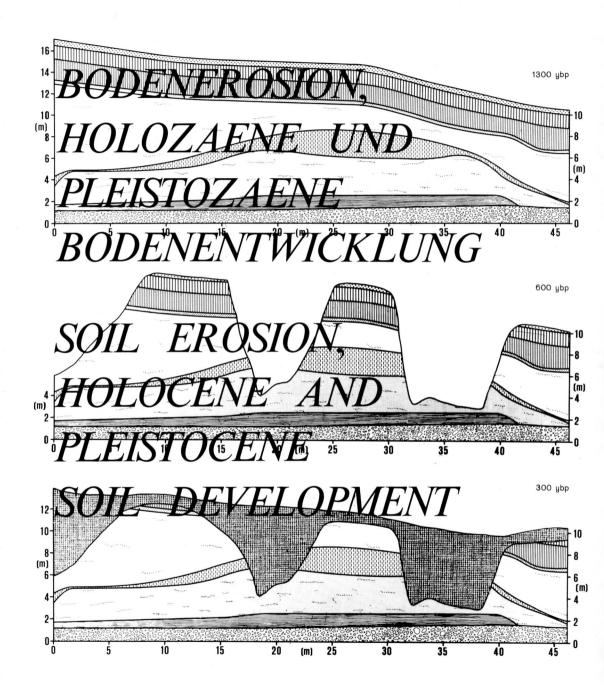

BODENEROSION, HOLOZAENE UND PLEISTOZAENE BODENENTWICKLUNG

SOIL EROSION, HOLOCENE AND PLEISTOCENE SOIL DEVELOPMENT

CATENA SUPPLEMENT 3

SLOPE RUNOFF AND SEDIMENT PRODUCTION IN TWO MEDITERRANEAN MOUNTAIN ENVIRONMENTS

M. **Sala**, Barcelona

Summary

Slope runoff and sediment production in the Montseny Mountain area of the Catalan Ranges are both slightly higher on shales than on granites. Runoff accounts for 10% of rainfall on granite and 13% on slates, while sediment production on grantie averages 152 mg-1 y-1. However, the shales yield many fragments greater than 2 mm diametre which combine with the fines to give a total yield of 203 gm-1 y-1. Variations in yield are high, both seasonally and from year to year. At the slate site, higher erosion rates were in autumn in two years and winter in one year, while at the granite site winter or spring/summer appeared to account for most erosion. Individual rain events can account for a large part of the total loss in any year. Processes in the shale locality are akin to those of a Mediterranean morphoclimatic environment, while those in the deeply weathered granitic locality resemble those of humid temperate regions. Local variations in denudation may thus reflect rock and geoecological conditions.

Resumen

La producción de escorrentía y de sedimento en la Montaña del Montseny, en las Cadenas Costeras Catalanas, tiene valores parecidos en el área de granitos y en la de pizarras. En el granito, sólo un 10% de la precipitación revierte en escorrentia mientras que en las pizarras la proporción es del 13%. La producción de sedimento es en el granito de 152 g m-1 a-1. Las pizarras producen muchos fragmentos de tamaño superior a los 2 mm de diámetro, lo cual, combinado con la proporción de material fino da un valor total de 203 g m-1 y-1. Las variaciones en la producción de sedimento son altas tanto estacionalmente como entre un año y otro. En el área de pizarras las tasas de erosión fueron más altas en otoño durante dos años y en invierno durante el tercer año, mientras que en el granito invierno y primavera aparecen como las estaciones más erosivas. En cualquiera de los años estudiados la mayor parte de la erosión se produce durante unos pocos de los acontecimientos lluviosos. Los procesos geomórficos en el área de pizarras son similares a los del medio morfoclimático mediterráneo, mientras que en al área de granito, que se encuentra profundamente meteorizado, tienen más similitud con los de las regiones templado-húmedas. Por tanto las

ISSN 0722-0723
ISBN 3-923381-12-3
©1988 by CATENA VERLAG,
D–3302 Cremlingen-Destedt, W. Germany
3-923381-12-3/88/5011851/US$ 2.00 + 0.25

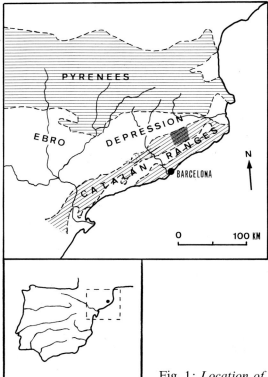

Fig. 1: *Location of study area.*

variaciones locales en el valor y tipo de denudación pueden ser el reflejo de las condiciones litológicas y geoecológicas.

1 Objectives of the Study

Since 1981 rainfall, runoff and sediment production have been studied in two areas of contrasted lithology in the Montseny Mountain. Despite many similar studies in humid environments, few were made in the Mediterranean environment until recently (RENDELL 1982, ROELS 1984, 1986, ROMERO-DIAZ et al. 1987, SALA 1981, 1984, VAN ASCH 1980).

Montseny mountain is the highest port (1714 m) of the centre of the Catalan Coastal Ranges, at the NE Iberian coast (fig.1) and comprises the two main lithologies, slates and granites, of the Hercynian part of the Ranges. Hence this massif to a certain extent represents the landforms and processes of a great part of the Ranges. Many Catalan naturalists work in the area, bringing the benefits of interdisciplinary cooperation, particularly at such field centres as La Castanya and Santa Fe.

2 Research Design

Field measurements were designed to obtain data that

(i) was representative of the general dynamics of these slopes (site A, one plot) and

(ii) indicated possible differences along the slopes (site B, three plots). See figures 2 and 6.

Each plot was instrumented with three GERLACH (1967) troughs inserted in different microtopographical positions and with 9 erosion pins inserted within one square metre. For a period one or two of the troughs were bounded at 2 m upslope in order to know precise water and sediment yield. Site A has one plot located approximately in the middle of the slope and site B has three plots located approximately in the upper, middle and lower parts of the slope.

Evidence of the importance of soil detachment by splash led to later installation of a splash collector in each plot. Similar observations of movement of rock fragments at the surface of the la Castanya field site led to the painting of stone lines within the bounded trough sreas so that subsequent movements could be measured.

3 Rainfall Characteristics: 1982–1984 The Study Period

Average annual rainfall for the 30 year record at the meteorological station of Turó de l'Home, at a 1700 m a.s.l. and at 7 km and 3 km distance from the research sites of La Castanya and Santa Fe respectively, is 1040 mm yr-1 of which 228 mm falls in winter, 250 mm in spring, 229 mm in summer, and 333 mm in autumn. October (137 mm), November (104 mm), September (102 mm), May (100 mm) and March (99 mm) are the rainiest months.

Annual totals during the study period were 1397 mm in 1982, 890 mm in 1983 and 1151 mm in 1984, a wet year being followed by a dry one while the third year had a little more than the average rainfall.

Wide seasonal contrasts exist, from 812 mm in the wet 1982 winter to 111 mm in spring. Generally autumn and winter are the wetter seasons. As much as 404 mm fall in a single month in November 1984, but some months receive less than 40 mm. Rain periods may be short but heavy, 368 mm (130 mm in one day) fell in six days in January 1982, while two periods of 3 and 5 days in February contributed a total of 309 mm, 168 mm falling in one day. High daily totals also occurred in October (141 mm) and on 14 March 1984 (182 mm).

4 La Castanya Field Station

The La Castanya field station (fig.2) on a NW slope of a third order tributary of La Mina, a headwater of the Tordera river, at 800 m a.s.l., has slope angles varying from 22° to 35°. Weathering and Quaternary periglacial processes on the slate substrate have produced a surficial mantle of rock debris wrapped in a silt and clay matrix up to 1 m thick in places. These unevenly distributed, variably sorted, heterogeneous mixture of coarse and fine sediments makes the hydraulic responses of slopes to rainfall action extremely uneven. Widespread rock outcrops, especially near the summits, add to the heterogenity of the slope surfaces.

The vegetation cover is a typical Mediterranean woodland of *Quercus ilex*, but unlike most green oak woodlands in the area has a poorly developed

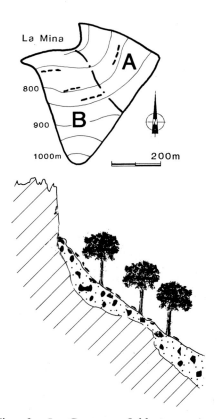

Fig. 2: *La Castanya field sites. a) research design b) slope morphology and vegetation cover.*

shrub and bush stratum. It is important to remember the fact that leaves are perennial, small and coriaceous, and litter formation is relatively slow.

4.1 Annual Rainfall, Runoff and Erosion Values

In the absence of a permanent raingauge, and following the observations made by RODÀ (1984), rainfall at La Castanya has been estimated to be 30% less than that at the meteorological station in Turó de l'Home. While the three year mean annual rainfall is 815 mm, yearly totals range from 1176 mm in 1982 to 600 mm in 1983 (tab.1a).

The collected runoff average was 104 l m-1 yr-1 and it varied annually from 180 l m-1 in the rainy year to 51 l m-1 during the dry year. Erosion was 203 g m-1 yr-1 and it ranged from 242 g m-1 to 129 g m-1. More than 80% of the sediment collected was rock fragments or aggregates of more than 2 mm diameter, and often bigger than 10 mm diameter. In the 1982 totals fragments bigger than 10 mm were not included because it was thought that their capture in the troughs was due to exprimental error.

If we assume water and sediment trapped in the open troughs (located approximately 100 m downslope of the divide) is the yield of this entire area, then the rates are lower as would be expected. But the results obtained in troughs bounded 2 m upslope were 27 l m-2 runoff and 100 g m-2 of sediment, of which only 15 g m-2 was material finer than 0.5 mm diametre (tab.1b). These values, although lower than the actual site measurements, are nevertheless much higher than the rates obtained extrapolating site measurements to the entire slope. We can conclude then that although the length of contributing area is sometimes greater than two metres (because absolute values obtained in the non-bounded troughs are higher than values obtained in the bounded ones), it is nevertheless under 100 metres, because the values obtained in the bounded trough are higher than those obtained if we consider runoff from open troughs coming from the total length of the slope. Hence in these slopes it appears that

year	rainfall (P) mm	runoff (Q) lm-1	fines (<2mm) gm-1	total erosion erosion (S) gm-1	(Q/P)	(S/P)	(S/Q)	(<2mm/S) %
1982	1176	180	53	242	.15	.20	1.34	22
1983	600	51	18	129	.09	.21	2.53	14
1984	670	80	46	239	.12	.35	2.99	19
Tot.	2446	311	117	610	.13	.25	1.96	19
Avg.	815	104	39	203	.13	.25	1.99	19

Tab. 1: a: La Castanya. Annual and average values of site measurements.

year	rainfall (P) lm-2	runoff (Q) lm-2	fines (<2mm) gm-2	total erosion erosion (S) gm-2	(Q/P)	(S/P)	(S/Q)	(<2mm/S) %
1982	1176	50	10	98	.04	.08	1.96	10
1983	600	11	13	56	.02	.09	5.09	23
1984	670	21	21	145	.03	.22	6.90	14
Tot.	2446	82	44	299	.03	.12	3.65	15
Avg.	815	27	15	100	3	.12	3.65	15

Tab. 1: b: La Castanya. Annual and average values obtained from a trough bounded 2 metres upslope.

there is not an important increase in runoff with increasing distance from the divide.

4.2 Seasonal Values

Winter and spring rainfall and runoff in each of the three years vary markedly, the first winter being extremely erosive when compared to the other two years, probably because rains were higher and more intense (fig.3). Save in the 1982 winter and 1984 summer, spring and autumn were generally erosive and winter and summer non erosive seasons.

4.3 Monthly and Weekly Values

The number of weeks in between measurements does not significantly effect the amount of rainfall, runoff and sediment collected; this is a typical characteristic of Mediterranean environments where often total monthly rainfall can arrive in a single week or in a couple of days (tab.2). Rainfall and runoff are highly correlated, while the closer association of erosion with rainfall than with runoff may indicate that volume and energy together are more important in denudation than runoff quantity, which is not significantly correlated with erosion (tab.2 and fig.4 and 5). Other processes related to rainfall and water in the slopes,

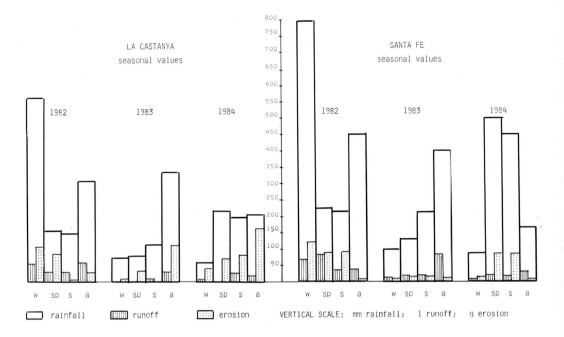

Fig. 3: *Seasonal values of rainfall, runoff and erosion in La Castanya and Santa Fe.*

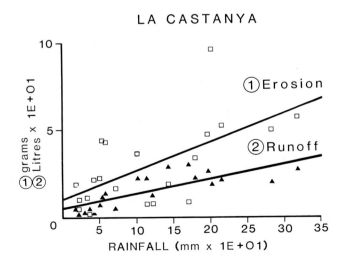

Fig. 4: *La Castanya. Rainfall, runoff and erosion relationships.*

Slope Runoff and Sediment Production

season	measurement interval days	rainfall mm	max. 24h rainfall mm	rainy days no.	runoff l/m	erosion g/m
Winter	34	318	91	14	27	58
Spring	37	282	93	13	20	50
Winter	64	215	128	20	21	53
Summer/Autumn	65	201	44	19	19	96
Spring/Summer	133	196	33	40	26	47
Winter	6	179	117	4	22	34
Autumn	40	170	99	7	30	9
Autumn	28	143	67	6	29	19
Summer/Autumn	115	121	16	34	13	8
Summer	68	114	53	14	23	7
Spring	14	101	25	8	22	36
Autumn/Winter	47	71	14	15	5	16
Winter	54	58	26	11	14	43
Spring	78	54	12	17	11	44
Autumn	25	50	25	5	8	23
Spring	49	43	8	12	3	21
Summer	13	36	17	6	5	1
Spring	48	33	10	6	1	11
Winter	20	23	10	3	5	10
Winter	9	21	6	5	1	5
Winter	19	18	12	3	5	19

	r.int.	runoff	erosion	<2 mm	2–10 mm	>10 mm
rainfall	r = .801 p = .001%	r = .791 p = .002%	r = .626 p = .2%	r = .653 p = .1%	r = .534 p = 1%	r = .437 p = 4.7%
r.int.		r = .745 p = .01%	r = .468 p = 3%	r = .482 p = 2.6%	r = .378 p = 9%*	r = .291 p = 9%*
runoff			r = .416 p = 6%*	r = .366 p = 10%	r = .379 p = 9%*	r = .233 p = 30%*
total erosion				r = .682 p = .06%	r = .973 p = .0001%	r = .544 p = 1%
sediment <2 mm					r = .431 p = 5%*	r = 309 p = 17%*
sediment 2–10 mm						r = .541 p = 1%

Tab. 2: *La Castanya*. Measurements sorted following rainfall magnitude. Correlation coefficients and level of significance. (+ not significantly different than 0).

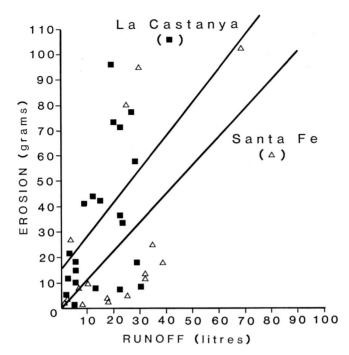

Fig. 5: *Runoff/erosion in La Castanya and in Santa Fe.*

such as wetting and drying, raindrop impact, soil humidity, are probably responsible for the mobilisation of debris, which constitutes the bulk of the eroded material on these slopes.

4.4 Granulometric Composition of Sediment Removed

Coarse material always dominates the sediment trapped, often accounting for more than 70% of the total. The porportions of coarse material collected in the bounded and in the open troughs do not differ significantly. Whether this similarity will persist is uncertain as the supply of coarse material from the bounded area may be exhausted. Small painted fragments (10 to 40 mm) within the bounded area have been observed to move downslope as far as 31 cm on a 27° slope in a single season (spring 1985), illustrating the role of surficial creep in the carriage of coarse material to the troughs.

4.5 Spatial Differences

At site A runoff to the open troughs 1 and 2 ranges from 86 m-1 and 44 l m-1 while the bounded trough 3 collects 32 l m-2 indicating the extend of natural spatial variability. Erosion rates in the three troughs are more similar, with rates of 80 g m-1, 64 g m-1 and 72 g m-2 in troughs 1, 2 and 3 respectively; the bounding of troughs appears to cause little difference in this three year study period. The proportion of coarse material differs widely from one trough to another at individual measurement times but annual averages are similar.

At site B the results obtained in the three plots installed along the slope are different. More water and sediment are trapped in the upper plot, located at the base of a rock outcrop, than in the lower one, located in a small flat area at the base of a slope. Here topography and slope material influence runoff and sediment production.

4.6 Pin and Washer Measurements

Observations by this technique show marked differences between individual measurements and in annual averages. In the first unusually rainy, study year average soil lowering for the nine pin 1 m-2 plots was as much as 34 mm yr-1 equivalent, at a soil density of 1.5 g cm-3, to an erosion rate of 51000 g m-2 yr-1. But the average for a 27 month period (November 1982 to January 1984) was 13 mm yr-1 lowering, yielding 19500 g m-2 yr. Despite the differences, rates are extremely high if compared with rates obtained with troughs. Pins were probably not long enough (30 cm) in comparison to the depth of regolith and so heave processes may partly explain these high values.

4.7 Splash Measurements

Observations of this process made after the main study period in the spring of 1985 cannot be correlated to the rest of the observations made in the preceeding three years but provide useful indicator. At site A the net splash detachment difference between up and down slope movement was 1457 g m-2. In site B similar rate was found in the upslope plot located near a rock outcrop, with 1523 g m-2 of soil movement. In the third slope plot the process was less active, with only 532 g m-2 particle detachment, and it was practically non existent in the plot at the bottom of the slope located in a flat surface usually covered by grass, with 71 g m-2 of soil detachment.

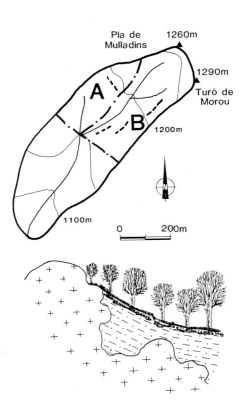

Fig. 6: *Santa Fe field sites: a) research design; b) slope morphology and vegetation over.*

5 Sante Fe Field Station

The site (fig.6) is located at the head waters of the Gualba drainage basin, at 1200 m a.s.l. and slope angles vary from 15° to 20°. Santa Fe is a more wet environment than La Castanya, with average annual rainfall exceeding 1000 mm and a good number of days with persistent mist. The underlying lithology is granitic (adamellite) which has been deeply weathered to a more than 15 m of gruss (CERVERA 1986). Further erosion of fine matrials has produced a typical tor and boulder landscape, with an alternation of tors in the interfluves, gruss and soil covered slopes, and boulder streams. Particle size distribution of slope materials is relatively homogeneous and consequently infiltration conditions can be expected to be fairly regular. The vegetation cover is typical of humid areas, consiting of a beech forest with little shrub or bush undercover. Seasonal changes, especially in the leaf dynamics, should be expected to have an influence on the slope processes.

5.1 Annual Rainfall, Runoff and Erosion Values

Average rainfall (Turó de l'Home) was 1247 mm and ranged from 1680 mm (1982) to 857 mm (1983); runoff was 128 l m-1 ranging from 186 l m-1 to 87 l m-1 and erosion was 152 gr m-1 with a maximum of 274 g m-1 and a minimum of 40 g m-1. Rates of runoff and erosion are low but if we relate plot measurements to the whole length of the slope (100 m) then values become extremely low. Contrarily, if we take rates from a trough bounded 2 m upslope, values are lower than average open plot values for runoff (63 l m-2) and fairly similar (143 g m-2) for erosion. Then, as at La Castanya field site, length of runoff can be greater than 2 metres but very much lower than the entire length of the slope (tab.3a and 3b).

5.2 Seasonal Values

Maximum rainfall occurs in autumn and winter while spring and summer receive a similar amount of rainfall. In all cases autumn appears as the less erosive season and summer and spring as the most erosive. Here, as in La Castanya, there seems to be some sort of alternation between and erosive season followed by a non-erosive one, although in an opposite manner to the alternation at La Castanya (fig.3). We could possibly assume that this difference is related to the different rhythmus between the perennial woodland of La Castanya and the deciduous woodland of Santa Fe and its implications for soil conditions.

5.3 Monthly and Weekly Values

As in La Castanya, weekly differences, especially when measurements were taken regularly, are extremely marked and show, as with seasonal observations, an alternance between erosive and non-erosive events.

The correlation is highly significant for runoff/rainfall and for runoff/rainfall intensity, significant for erosion/runoff and erosion/rainfall intensity and low for erosion/rainfall (tab.4, fig.5 and 7). Unlike at La Castanya site this seems an environment strongly controled by slope hydrology.

5.4 Granulometric Composition of Sediment

Sediment smaller than 2 mm diameter always predominate, with annual averages of 72% of the total, and is never biger than 10 mm. This probably explains the good correlation between runoff and erosion due to the fact that particles are available for transport by overland flow.

5.5 Spatial Differences

Microtopographic location of troughs produces marked differences in runoff and sediment collection in the same site during a single observation (6 to 22 l

year	rainfall (P) mm	runoff (Q) l/m	fines (<2mm) g/m	total erosion erosion (S) g/m	(Q/P)	(S/P)	(S/Q)	(<2mm/S) %
1982	1680	186	169	274	.11	.16	1.47	62
1983	857	87	33	40	.10	.05	.46	83
1984	1204	110	128	143	.09	.12	1.30	90
Tot.	3742	383	330	457	.10	.11	1.19	72
Avg.	1247	128	110	152	.10	.11	1.19	72

Tab. 3: *a: Santa Fe. Annual and average values of site measurements.*

year	rainfall (P) mm	runoff (Q) lm-2	fines (<2mm) gm-2	total erosion erosion (S) gm-2	(Q/P)	(S/P)	(S/Q)	(<2mm/S) %
1982	1680	90	153	239	.05	.14	2.66	64
1983	857	31	13	20	.04	.02	.65	65
1984	1204	68	157	171	.06	.14	2.51	92
Tot.	3742	189	323	430	.05	.11	2.27	75
Avg.	1247	63	108	143	.05	.11	2.27	75

Tab. 3: *b: Santa Fe. Annual and average values for a bounded trough collecting runoff and stemflow.*

m-1 in runoff and 1 to 24 g m-1). In contrast to La Castanya annual averages are quite similar for runoff (10 l m-1 to 8 l m-1) and vary for erosion (22 g m- to 5 g m-1).

Individual trough measurements at site A show similar runoff values but different values for erosion. Correlation for rainfall and runoff are good for the three samples, but values vary from trough to trough for runoff and erosion and for rainfall and erosion.

Observations from the three plots of site B, which are located at different positions along the slope, show higher runoff and erosion at the upper part while differences between the two lower sites are not remarkable. This is thought to be caused by the presence of tors in the interfluves producing areas of higher runoff and of sediment entrainment. A few measurements made very near the talweg showed flooding during strong rains.

5.6 Pin and Washer Measurements

An average of 10 mm yr-1 lowering has been measured, with two years of erosion and one year accumulation (1983). Balance for the three years is -29.8 mm. If we take soil density as 1.5 g cm-3 this would mean 1500 g m-2 yr-1 erosion, a high value if compared to that estimated from trough measurements (2.5 mm yr-

season	measurement interval days	rainfall mm	rainfall int. 24h mm	rainy days no.	runoff lm-1	erosion gm-1
Winter/Spring/Summer	267	952	182	87	68	114
Winter	34	454	130	14	38	17
Spring	37	403	133	13	36	11
Spring/Summer	227	353	35	57	34	25
Winter	7	255	168	4	27	81
Autumn	49	242	141	7	26	6
Autumn	27	204	95	6	17	4
Autumn	38	165	32	13	32	13
Summer	53	161	76	13	30	95
Spring	13	144	36	7	25	29
Autumn/Winter	105	101	20	15	18	4
Winter	52	83	37	11	10	9
Spring	63	61	17	12	8	8
Summer	14	51	24	6	8	2
Winter	19	33	14	3	1	3
Winter	8	30	9	5	1	0
Winter	19	26	17	3	3	26
Spring	27	18	5	6	1	4

	season	rainfall	r.int.	raindays	runoff	erosion
season		r = .807 p = .003%	r = .498 p = 3%	r = .818 p = .002%	r = .685 p = .1%	r = .591 p = 8%
rainfall			r = .793 p = .005%	r = .830 p = .001%	r = .920 p = .001%	r = .627 p = .4%
r.int.				r = .391 p = .01%	r = .763 p = .05%	r = .593 p = .7%
raindays					r = .761 p = .02%	r = .566 p = .1%
runoff						r = .678 p = .1%

Tab. 4: *Santa Fe. Measurements sorted using rainfall magnitude. Correlation coefficients and level of significance.*

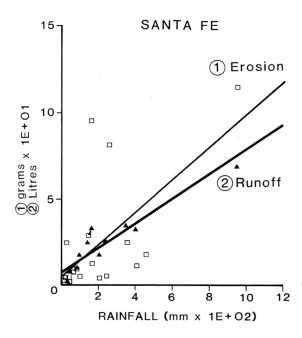

Fig. 7: *Santa Fe: rainfall, runoff and erosion.*

1). Nevertheless it has to be taken into account that this is an environment with a great deal of litter and decomposed organic matter, making pin measurements difficult.

Observations of material adhering to the pins suggested splash may be an important process especially after leaf fall, but no quantitative estimates are available to confirm this.

6 Discussion and Results

Although average values of runoff and sediment removal in the slate and in the granite landscapes of Montseny mountain are fairly similar, they reflect different processes of slope evolution. The most remarkable feature in the slope dynamics of La Castanya is the predominance of the movement of coarse material by surface creep leading to high erosion rates of these slopes. In Santa Fe the main geomorphic process is slope wash and the entrainement of fine soil particles. Consequently, runoff and erosion are better correlated at Santa Fe than in La Castanya, where significant correlation is only between rainfall and erosion.

Average annual values for sediment removal are moderate as is common on vegetated slopes, and do not differ in order of magnitude in the two studied enviroments. Results obtained are within the range of data from other Mediterranean environments YOUNG (1974) including Southern France (GABERT 1964, BIROT 1970, VAN ASCH 1980, ROELS 1984) and California (KRAMMES 1963). They are also similar to observations from temperate maritime (IMESON et al. 1980) and subtropical and tropical environments (WILLIAMS 1973), as compiled by SAUNDERS & YOUNG (1983). As for runoff, the low proportion of water collected as overland flow indicates that infiltration is high in both environments. Studies in progress in La Castanya (BONILLA, personal communication) seem to indicate that a great deal of water is taken by tree roots or is directed to the bedrock layer because subsurface flow is also low, while recent survey in Santa Fe (CERVERA, personal communication) points to the relative importance of subsurface flow.

Seasonality is marked and presents an alteration between erosive and non-erosive periods, but the trends vary in the two sites, both in runoff and in sediment removal. As fundamental climatic variations cannot be claimed, the different responses to rainfall, both in the generation of overland flow and of erosion, can only be explained by basic soil and vegetation characteristics and their conditions immediately before, during and after the rainfall events. The debris slopes of La Castanya behave similarly to slopes in a dry-mediterranean climate, where effectiveness of storms is greatest in spring and autumn (THORNES 1980). In Santa Fe the system functions like a humid enviroment (DUNNE 1978, ANDERSON & BURT 1978) because of higher water retention of its thicker regolith and soils throughout the seasons, provoking a regular overland flow response to rainfall and consequent effectiveness in the washing of available material during winter frost action and summer wetting and drying.

Weekly observations exhibit maximal variations, often with a marked alternation between high and low erosion rates, a fact that adds to the importance given to sediment availability as pointed out by several authors (KIRKBY 1980, ROELS 1984) and in subsequent research in Santa Fe (CERVERA 1986).

Spatial differences within each site are marked and they occur both along the slopes and from plot to plot. Along the slopes runoff and erosion appear to be more closely related to surface properties and to its corresponding response to hydraulic processes than to position along the slope. Microtopography and vegetation seem clearly responsible for differences between plots. In addition, for a given plot, variations in the percentage of runoff and sediment production differ between measurements, although average results seem consistent. Here again antecedent soil and vegetation conditions seem to be the cause of some differences. Both in La Castanya and in Santa Fe differences in runoff are clearly related to the rock/soil ratio, as used by YAIR & LAVEE (1985), although the downslope behaviour differs at each site, with dry footslopes in La Castanya, as in YAIR & LAVEE's model (1974), and a tendency to saturation near the slope foot stream in Santa Fe, as it was reported in humit environments (DUNNE & BLACK 1970, ANDERSON & BURT 1978). We can thus classify the slate debris slopes as "dry" and the granite grus slopes as "wet". In relation to erosion, "sediment avaiability" seems to be important irrespective of "water availability".

Correlations between rainfall and runoff are better in open than in bounded troughs, the highest being for an open trough in Santa Fe locatd in a concavity. Runoff/erosion correlations are not very clear, but in any case they do not seem related to the bounding of troughs. Findings by VAN ASCH (1980) and ROELS (1984, 1985), which indicate that the length of flow contributing to the trough is generally around 2.5 m and seldom longer than 8 m seems also to apply at Montseny. The length of sediment contributing area is certainly no more than 2 m, which accords with the point to point transfer of sediment as state by VAN ASCH (1980).

The importance of splash became obvious when sediment was regularly found attached to fallen leaves and to erosion pins, especially in the decideous forest of Santa Fe. Later measurements showed that splash was also important in the green oak woodland of La Castanya in spite of its debris paved slopes. Splash processes under deciduous forest have been pointed out by several authors including IMESON (1977) and KWAAD (1977) and also the significance of sediment transport through litter (VAN ZON 1978). In semi-arid environments rainfall impact can also have a positive effect on the movement of debris, as observed by SCHUMM (1967). Together with runoff induced creep (DE PLOEY & MOEYERSONS 1975), splash must be the most active processes in the evolution of the slopes of La Castanya.

The role of vegetation in the production of overland flow and erosion was clearly observed under the beech forest of Santa Fe, where in trough 1, located near a tree, the most extreme and discontinuous events were measured. RODÁ (1984) has estimated stemflow in Santa Fe as 10–14% of total rainfall and 8–10% in La Castanya.

7 Conclusions

Parent rock and its weathering products create two denudational environments within the Montseny area, one dominated by the slow mass movement of debris, and the other by the flow of water on the slopes. With average rainfall above 800 mm the rock and debris slopes of La Castanya behave rather similarly to semi-arid slopes, while the grus and boulder slopes of Santa Fe have a response that is more characteristic of humid environments.

Variability in all aspects of erosion is outstanding and these provide examples of what DE PLOEY (1981) terms ambivalent effect of erosion, and in addition they reflect hydrological and hydraulic soil properties (KIRKBY 1980).

Acknowledgement

This research was initiated with a grant from FUNDACION JUAN MARCH (1981) and later funding has been provided by CIRIT. We are also grateful for facilities provided by the field centres of El Vilar de la Castanya (Servei de Protecció de la Natura, Direcció General del Medi Rural, Generalitat de Catalunya) and of Santa Fe del Montseny (Servei de Parcs Naturals, Diputació de Barcelona). Visiting colleagues to the field sites have provided invaluable ideas related to research design and interpretation of results, for which we are deeply indebted. And finally my gratitude for the help of the N. Cox, Th. van Ash and the anonymous referees whose corrections and ideas have greatly improved the first draft of this paper.

References

ANDERSON, M.G. & BURT, T.P. (1978): Experimental investigation concerning the topographical control of soil water movement on hillslopes. Zeitschrift für Geomorphologie Supp. **29**, 52–63.

BIROT, P. (1970): Etude qantitatif des processus érosifs agissant sur les versants. Zeitschrift für Geomorphologie Supp. **9**, 10–43.

BRYAN, R.B. (1979): The influence of slope angle on soil entrainment by sheetwash and rainsplash. Earth Surface Processes, **4**, 43–58.

BRYAN, R.B. (1981): Soil erosion under simulated rainfall in the field and laboratory; variability of erosion under controlled conditions. Proceedings of the IAHS Symposium on Erosion and Sediment Transport Measurement, 391–404.

CARSON, M.A. & KIRKBY, M.J. (1972): Hillslope form and process. Cambridge University Press.

CERVERA, M. (1986): Spatial variation of surface wash and erosion in the slopes of Santa Fe. In: Sala, M., Gallart, F. & Clotet, N. (eds.), Excursion Guide Book, COMTAG SYMPOSIUM, 69–72.

DE PLOEY, J. (1981): The ambivalent effects of some factors of erosion. Mém. Inst. Geol. Univ. Louvain, XXXI, 171–181.

DE PLOEY, J. & MOEYERSONS, J. (1975): Runoff creep of coarse debris: experimental data and some field observations. CATENA, 2, 275–288.

DUNNE, T. (1978): Field studies of hillslope flow processes. In: Kirkby, M.J. (ed.). Hillslope hydrology, Wiley, 227–290.

DUNNE, T. & BLACK, R.D. (1970): Partial area contribution to storm runoff in a small New England watershed. Water Resources Research, 6, 1269–1311.

GABERT, P. (1964): Premiers resultats de mesures d'érosion sur des parcelles expérimentales dans la région d'Aix-en-Provence (Bouches du Rhône-France). Zeitschrift für Geomorphologie Supp. 5, 213–214.

GERLACH, T. (1967): Hillslope troughs for measuring sediment movement. Révue de Géomorphologie Dynamique, 4, 173–174.

IMESON, A. (1977): Splash erosion, animal activity and sediment supply in a small forested Luxembourg catchment. Earth Surface Processes, 2, 153–160.

IMESON, A. & KWAAD, F.J.P.M. (1976): Some effects of burrowing animals on slope processes in the Luxembourg Ardennes. Geografiska Annaler, 58 A, 317–328.

KIRKBY, M.J. (1980): The problem. In: Kirkby & Morgan (eds.). Soil erosion. Wiley, 1–16.

KRAMESS, J.S. (1963): Seasonal debris movement from steep mountainside slopes in Southern California. Misc. Publs. U.S. Dep. Agric., 970, 85–88.

KWAAD, F.J. (1977): Measurements of rainsplash erosion and the formation of colluvium beneath deciduous woodland in the Luxembourg Ardennes. Earth Surface Processes, 2, 161–174.

MEYER, L.D., FOSTER, G.R. & ROMKENS, M.J.M. (1975): Source of soil eroded by water from upland slopes. USDA ARS-S-40, 177–189.

MORGAN, R.P.C. (1980): Field studies on sediment transport by overland flow. Earth Surface Processes, 5, 307–317.

RENDELL, H. (1982): Clay hillslope erosion rates in the basento valley, S. Italy. Geografiska Annaler, 64 A, 141–147.

ROMERO-DIAZ, A., LOPEZ-BERMUDEZ, F., THORNES, J., FRANCIS, C. & FISHER, G. (1987): Overland flow erosion rates in a semi-arid Mediterranean environment for an experimental site, Murcia, Spain. In: Harvey, A. & Sala, M. (eds.), Geomorphic processes in environments with strong seasonal contrasts. CATENA SUPP. 12.

RODA, F. (1984): Economia hidrica. In: Terradas (ed.). Introdució a l'ecologia del faig al Montseny. Diputació de Barcelona, 59–72.

ROELS, J.M. (1984): Surface runoff and sediment yield in the Ardèche rangelands. Earth Surface Processes and Landforms, 9, 371–381.

ROELS, J.M. (1985): Estimation of soil loss at a regional scale based on plot measurements — Some critical considerations. Earth Surface Processes and Landforms, 10, 587–595.

SALA, M. (1981): Geomorphic processes in a small Mediterranean drainage basin (Catalan Ranges). Transactions Japanese Geomorphological Union, 2, 239–251.

SALA, M. (1983): Fluvial and slope processes in the Fuirosos basin, Catalan Ranges, north east Iberian coast. Zeitschrift für geomorphologie, 27, 393–411.

SAUNDERS, I. & YOUNG, A. (1983): Rates of surface processes on slopes, slope retreat and denudation. Earth Surface Processes and Landforms, 8, 473–501.

SCHUMM, S.A. (1964): Seasonal variations of erosion rates and processes on hillslopes in western Colorado. Zeitschrift für Geomorphologie Supp. 5, 215–238.

SCHUMM, S.A. (1967): Rates of surficial rock creep on hillslopes in western Colorado. Science. N.Y. 55, 560–561.

THORNES, J. (1980): Erosional processes of running water and their spatial and temporal controls: a theoretical viewpoint. In: Kirkby & Morgan (eds.). Soil erosion. Wiley, 129–182.

VAN ASCH, Th.W.J. (1980): Water erosion on slopes and landsliding in a Mediterranean landscape. Utrechtss Geografische Studies **20**, Geografisch Instituut Rijksuniversiteit Utrecht, 238 pp.

VAN ZON, H.J. (1978): Litter transport as a geomorphic process. Publ. Fys. Geogr. Bodemkundig Lab. Univ. Amsterdam, **24**, 134 pp.

YAIR, A. & LAVEE, H. (1974): Areal contribution to runoff on scree slopes in an extreme arid environment: a simulated rainstorm experiment. Zeitschrift für geomorphologie Supp. **21**, 106–121.

YAIR, A. & LAVEE, H. (1985): Runoff generation in arid and semi-arid zones. In: Anderson & burt (eds.). Hydrological Forecasting, Wiley, 183–220.

YOUNG, A. (1974): The rate of slope retreat. In: Brown & Waters (eds.). Progress in Geomorphology. Inst. British Geogr., 65–78.

Address of author:
M. Sala
Departament de Geografia Fisica
i Analisi Regional
Universitat de Barcelona
08028 Barcelona, Spain

Jan de Ploey (Ed.)

RAINFALL SIMULATION, RUNOFF and SOIL EROSION

CATENA SUPPLEMENT 4, 1983

Price: DM 120,-

ISSN 0722-0723 ISBN 3-923381-03-4

This CATENA-Supplement may be an illustration of present-day efforts made by geomorphologists to promote soil erosion studies by refined methods and new conceptual approaches. On one side it is clear that we still need much more information about erosion systems which are characteristic for specific geographical areas and ecological units. With respect to this objective the reader will find in this volume an important contribution to the knowledge of active soil erosion, especially in typical sites in the Mediterranean belt, where soil degradation is very acute. On the other hand a set of papers is presented which enlighten the important role of laboratory research in the fundamental parametric investigation of processes, i.e. erosion by rain. This is in line with the progressing integration of field and laboratory studies, which is stimulated by more frequent feed-back operations. Finally we want to draw attention to the work of a restricted number of authors who are engaged in the difficult elaboration of pure theoretical models which may pollinate empirical research, by providing new concepts to be tested. Therefore, the fairly extensive publication of two papers by CULLING on soil creep mechanisms, whereby the basic force-resistance problem of erosion is discussed at the level of the individual particles.

All the other contributions are focused mainly on the processes of erosion by rain. The use of rainfall simulators is very common nowadays. But investigators are not always able to produce full fall velocity of waterdrops. EPEMA & RIEZEBOS give complementary information on the erosivity of simulators with restricted fall heights. MOEYERSONS discusses splash erosion under oblique rain, produced with his newly-built S.T.O.R.M-1 simulator This important contribution may stimulate further investigations on the nearly unknown effects of oblique rain. BRYAN & DE PLOEY examined the comparability of erodibility measurements in two laboratories with different experimental set-ups They obtained a similar gross ranking of Canadian and Belgian topsoils.

Both saturation overland flow and subsurface flow are important runoff sources under the rainforests of northeastern Queensland. Interesting, there, is the correlation between soil colour and hydraulic conductivity observed by BONELL, GILMOUR & CASSELLS. Runoff generation was also a main topic of IMESON's research in northern Morocco, stressing the mechanisms of surface crusting on clayish topsoils.

For southeastern Spain THORNES & GILMAN discuss the applicability of erosion models based on fairly simple equations of the "Musgrave-type" After Richter (Germany) and Vogt (France) it is TROPEANO who completes the image of erosion hazards in European vineyards. He shows that denudation is at the minimum in old vineyards, cultivated with manual tools only. Also in Italy VAN ASCH collected important data about splash erosion and rainwash on Calabrian soils. He points out a fundamental distinction between transport limited and detachment-limited erosion rates on cultivated fields and fallow land. For a representative first order catchment in Central–Java VAN DER LINDEN comments contrasting denudation rates derived from erosion plot data and river load measurements. Here too, on some slopes, detachment-limited erosion seems to occur

The effects of oblique rain, time-dependent phenomena such as crusting and runoff generation, detachment-limited and transport-limited erosion including colluvial deposition, are all aspects of single rainstorms and short rainy periods for which particular, predictive models have to be built. Moreover, it is argued that flume experiments may be an economic way to establish gross erodibility classifications. The present volume may give an impetus to further investigations and to the evaluation of the proposed conclusions and suggestions

Jan de Ploey

G.F. EPEMA & H.Th. RIEZEBOS
FALL VELOCITY OF WATERDROPS AT DIFFERENT HEIGHTS AS A FACTOR INFLUENCING EROSIVITY OF SIMULATED RAIN

J. MOEYERSONS
MEASUREMENTS OF SPLASH–SALTATION FLUXES UNDER OBLIQUE RAIN

R.B. BRYAN & J. DE PLOEY
COMPARABILITY OF SOIL EROSION MEASUREMENTS WITH DIFFERENT LABORATORY RAINFALL SIMULATORS

M. BONELL, D.A. GILMOUR & D.S. CASSELLS
A PRELIMINARY SURVEY OF THE HYDRAULIC PROPERTIES OF RAINFOREST SOILS IN TROPICAL NORTH–EAST QUEENSLAND AND THEIR IMPLICATIONS FOR THE RUNOFF PROCESS

A.C. IMESON
STUDIES OF EROSION THRESHOLDS IN SEMI–ARID AREAS: FIELD MEASUREMENTS OF SOIL LOSS AND INFILTRATION IN NORTHERN MOROCCO

J.B. THORNES & A. GILMAN
POTENTIAL AND ACTUAL EROSION AROUND ARCHAEOLOGICAL SITES IN SOUTH EAST SPAIN

D. TROPEANO
SOIL EROSION ON VINEYARDS IN THE TERTIARY PIEDMONTESE BASIN (NORTHWESTERN ITALY): STUDIES ON EXPERIMENTAL AREAS

TH.W.J. VAN ASCH
WATER EROSION ON SLOPES IN SOME LAND UNITS IN A MEDITERRANEAN AREA

P VAN DER LINDEN
SOIL EROSION IN CENTRAL–JAVA (INDONESIA). A COMPARATIVE STUDY OF EROSION RATES OBTAINED BY EROSION PLOTS AND CATCHMENT DISCHARGES

W.E.H. CULLING
SLOW PARTICULARATE FLOW IN CONDENSED MEDIA AS AN ESCAPE MECHANISM: I. MEAN TRANSLATION DISTANCE

W.E.H. CULLING
RATE PROCESS THEORY OF GEOMORPHIC SOIL CREEP

SOIL ORGANIC HORIZONS OF MEDITERRANEAN FOREST SOILS IN NE-CATALONIA (SPAIN): THEIR CHARACTERISTICS AND SIGNIFICANCE FOR HILLSLOPE RUNOFF, AND EFFECTS OF MANAGEMENT AND FIRE

J. **Sevink**, Amsterdam

Summary

In research thusfar carried out little attention has been paid to the topsoil characteristics of mediterranean forest soils and their effects on hill slope runoff. These soils were assumed to have a mull type humus, due to rapid decomposition of the litter, but this theory is based on limited evidence. Here, first results are presented of a study on the organic horizons of forest soils and their significance for hillslope runoff, carried out in NE Catalonia, an area with an attenuated mediterranean climate and with non-calcareous parent materials.

General trends in soil development were studied in two altitudinal sequences, on granodiorite and on schist respectively. Small plots with a (semi)natural vegetation were studied for their profile development and representative profiles were analysed for their water-soluble elements and reaction.

In the mediterranean area soils under (semi)natural forest on granodiorite have a thinner ectorganic profile (L + F + H), but a thicker H horizon than the soils on schist. However, both types of soils exhibit a distinct accumulation of organic matter, most probably due to a strong reduction of the rate of litter decomposition during its later stages. Nutrient concentrations in the ectorganic profiles are high, suggesting that during the first stages decomposition rates are high. Differences in concentrations of individual elements, which are only slight, seem to be primarily controlled by the parent material.

Soils under planted pine forests and chestnut forests have ectorganic profiles with deviating characteristics: relatively thin and low in nutrients under pine, and relatively thick and very high in nutrients under chestnut. These differences are attributed to differences in litter input and litter composition.

In the boreo-alpine area, mainly under beech forest, thicknesses of ectorganic profiles are considerable. They are strongly related to vegetation, but probably also to parent material and exposition. Nutrient concentrations and re-

ISSN 0722-0723
ISBN 3-923381-12-3
©1988 by CATENA VERLAG,
D–3302 Cremlingen-Destedt, W. Germany
3-923381-12-3/88/5011851/US$ 2.00 + 0.25

action differ markedly from those in the mediterranean zone.

Specific research, subsequently carried out in the mediterranean area, concentrated on the microvariability of ectorganic profiles in forest soils and on the effects of fire and forest management on these profiles, in particular on their infiltration characteristics.

Microvariability was found to be related to local variation in species-dependent litter input and composition, as well as in litter transport by surface runoff.

Fire was found to lead to a nearly complete (if intense) or partial (if less intense) destruction of the ectorganic horizons. Similar effects are produced by removal of the undergrowth, as a result of increased erosion.

Rainfall simulator experiments showed that under (semi)natural vegetation hill slope runoff will probably only occur during heavy rainstorms and is largely determined, at least during the summer period, by the combination of an ectorganic profile with high storage and infiltration capacity and a waterrepellent upper part of the mineral soil. Solute concentrations in this runoff are probably high.

Destruction of the ectorganic profile by fire leads to a strong increase in hill slope runoff, probably with initially high, but in time low solute contents. Similar, but less strongly expressed changes in quantity and composition of hill slope runoff will occur as a result of the removal of the undergrowth.

Resumen

Se han estudiado las tendencias generales del desarrollo de los suelos en dos secuencias altitudinales sobre granodiorita y sobre esquistos. El trabajo se ha llevado a cabo en pequeñas parcelas con vegetación (semi)natural, en las que se ha estudiado el desarrollo de sus perfiles y se han analizado los elementos solubles y su reacción.

En las áreas mediterráneas los suelos que se encuentran bajo bosque (semi)natural y sobre granodiorita tienen un perfil ectorgánico más fino (L + F + H) pero un horizonte H más grueso que los suelos sobre esquisto. No obtante, ambos tipos de suelos presentan una acumulación distinta de materia orgánica, probablemente debido a la fuerte reducción de la tasa de descomposición del mantillo durante sus últimos estadios. Las concentraciones de nutrientes en los perfiels ectorgánicos son altos, lo que lleva a pensar que durante los primeros estadios de descomposición las tasas son altas. Las diferencias en concentraciones de los elementos individuales son débiles y parecen estar primordialmente controlados por la roca madre.

Los perfiles ectorgánicos de los suelos de los bosques de pinos son relativamente delgados y con un contenido bajo en nutrientes, mientras que en los bosques de castaños son relativamente gruesos y con un contenido alto en nutrientes; estas diferencias hay que atribuirlas a las diferencias en la cantidad y composición del mantillo.

También se ha encontrado que el fuego es causa de una destrucción más o menos completa de los horizontes ectorgánicos y que se produce una destrucción similar con la tala del sotobosque a causa de un incremento de la erosión.

Experimentos llevados a cabo un simulador de lluvia muestran que bajo una cobertera vegetal (semi)natural el escurrimiento de agua por las vertientes probablemente solo ocurre en el caso de lluvias intensas, y está en gran parte determiando, por lo menos durante el período estival, por la combinación de un perfil ectorgánico con una gran capacidad de infiltración y almacenaje y una parte superior del suelo mineral que repele el agua.

1 Introduction

Throughout the Mediterranean area, including those parts with a more attenuated climate, accelerated soil erosion is considered as a serious and often aggravating problem. It is generally attributed to changes in hill slope runoff characteristics, resulting from changes in the vegetational cover and in the top soil characteristics brought about by man, mainly through intensive agriculture, deforestation and ever increasing forest and shrub fires. It leads to losses of nutrients and is one of the major processes causing land degradation.

In so far as the more attenuated parts of the Mediterranean area are concerned, very little attention has been paid to the characteristics of and processes in topsoils under natural vegetation and to the effects of deforestation and fires on these characteristics and processes. Studies by forest ecologists (see a.o. DI CASTRI et al. 1973 and KRUGER et al. 1983: Mediterranean type ecosystems) have been mainly focussed on nutrient cycling in and nutrient status of shallow soils over limestone or other highly calcareous parent materials under macchia or garrigue and to the effects of forest fires on these soil characteristics. Nearly all studies by pedologists deal with similar soils or with agricultural soils. Results from these studies have led to the rather widely held theory (see DUCHAUFOUR 1982) that, under the mediterranean climatic conditions described above, accumulation of organic matter plays a subordinate role, due to rapid mineralization of the litter input, and that topsoils have a mull type humus.

Recent studies by the Laboratory of Physical Geography and Soil Science, University of Amsterdam, in Central Italy (VOS & PEDROLI 1985, PEDROLI et al. 1985, V.BREMEN 1985, 1986) indicate that, other than generally assumed in forest soils on non-calcareous parent materials accumulation of organic matter plays an important role. Clear evidence was found for a strong influence of vegetation on the characteristics of and processes in the organic and mineral horizons.

Studies on the general trends in the development of organic soil horizons under forest, carried out in Catalonia, NE-Spain (Montseny and adjacent Selva area), confirmed these observations. Additional, more specific studies in the same area showed that hill slope runoff characteristics are strongly influenced by the organic soil horizons and that the impact of land use and of forest fires is considerable. Results of the studies in Catalonia, which in part are rather preliminary, are presented in this paper.

2 General Information

The area ivestigated (see fig.1) has a geology dominated by acid igneous and metamorphic rocks and a rather strong relief, some summits reaching an altitude of about 1700 m. As a result of the latter large variations in climate occur as well as a large variation in vegetation types, ranging from distinctly mediteranean (Quercetum ilicis galloprovinciale

Fig. 1: *Location of the area.*

suberetosum) to boreo-subalpine (SALA 1979, BOLOS 1983). Some climatic data are presented in tab.1.

In particular in the mediterranean part of the area large differences in human impact on the vegetation exist, connected with local differences in landuse. Forest fires in this relatively dry part are a common phenomenon, but until yet have not resulted in the complete destruction of older well developed forests. Thus a whole range from virtually natural vegetation, through more or less intensively exploited vegetation (such as *Quercus suber* and *Fagus sylvatica* forests), to monocultural forest plantations (*Castanea sativa, Pinus* species or *Eucalyptus*) can be found, as well as vegetations regenerating after fire damage.

3 Methods

To rapidly identify general trends in profile development, related to differences in climate, parent material and forest type, a preliminary study of a series of plots (18) was carried out. Criteria used for their selection were exposition (N and S), altitude and parent material (schist and granodiorite). Height intervals between the plots were about 200 metres and care was taken to locate plots in forests, which were as natural as possible. Plot sizes were about 50 × 200 m and within these plots at least 15 profiles, located on sample lines parallel to the slope, were described. Particular attention was paid to the organic matter profile (or humus form profile, see KLINKA et al. 1981).

Based on the information obtained by this inventory study, a next series of plots were selected, in which the major trends in profile development could be studied and in which the vegetation had been minimally affected by man or fire. In addition, within the mediterranean zone, two additional plots under forest plantations were selected to study the effect of vegetation. General information on these plots is given in tab.2. Plots sizes were about 30 × 50 m and all plots were situated on straight to nearly straight, sloping to moderately steep, non-colluvial slopes in larger, uniform forest complexes.

Within each plot 10 soil profiles were described for both their mineral and organic horizons. The first profile was choosen at random. The next 4 profiles were on a sample line, parallel to the slope, through the first profile. The other 5 profiles were on a parallel sample line, at a distance of 20 m. Intervals between profiles on these lines were 10 m. Organic horizons are noted for their spatial microvariability and therefore another 40 profiles were described for their organic horizons only. These were located at a distance of 2.0 m from the above mentioned profiles (4 per each profile, on perpendicular sample lines).

	J	F	M	A	M	J	J	A	S	O	N	D	Year
Balenya	3.5	5.7	9.8	11.2	15.2	18.4	20.8	20.5	18.7	14.3	7.4	5.1	12.5°C
Santa Fe	1.0	3.3	6.5	7.6	10.9	13.7	16.6	16.7	14.0	10.9	6.4	2.3	9.1°C
Turo de l'Home	-0.4	0.8	1.8	5.3	8.2	12.0	15.4	13.7	11.6	7.3	3.6	1.0	6.5°C
Montseny, poble	43.1	76.6	86.8	81.0	107.5	65.5	35.8	43.9	76.2	89.0	89.0	65.2	859.7 mm
Viladrau	40.6	86.5	98.8	85.6	98.8	72.3	39.0	53.3	89.6	104.0	108.6	82.9	960.0 mm
Turo de l'Home	58.9	68.3	99.0	83.9	101.2	72.0	42.8	83.3	108.4	125.2	95.0	107.5	1045.5 mm
Granollers	36.2	41.3	58.2	54.1	62.8	42.3	29.5	49.0	72.6	65.4	48.5	52.7	612.9 mm

Altitudes above sea level: Balenya: 580 m; Santa Fe: 1130 m; Turo de l'Home 1707 m; Montseny, poble: 552 m; Granollers: 140 m

Tab. 1: *Climatic data from stations, representative for the various climatic zones.*

Site number	Parent material	Exposition	Altitude	Vegetation (major species)
5	Granodiorite	North	400 m	*Quercus sub.* + *il.*, *Arbutus unedo*, *Erica arborea*
14			500 m	*Castanea sativa*
13			600 m	*Pinus*
1			1200 m	*Fagus*
4		South	400 m	*Quercus sub.* + *il.*
8			1200 m	*Fagus*
11	Schists	North	200 m	*Quercus sub.* + *il*
7			1150 m	*Fagus*
6			1200 m	*Fagus, Juniperus*
2			1700 m	Grasses
12		South	200 m	*Quercus suber*
9			950 m	*Quercus il.*
10			1300 m	*Erica, Juniperus*, Ferns
3			1700 m	Grasses

Tab. 2: *Characteristics of the plots studied.*

The profile data were statistically analyzed (mean thicknesses of individual horizons and standard deviations). The results were used to identify "characteristic profiles", i.e. profiles of which thicknesses of individual horizons are close to the mean, within the group of profiles studied. These were subsequently sampled for further analyses. Both description and sampling were carried out in summer 1985.

For the description of the mineral soil profile the Guidelines of the FAO (1977) were used. A slightly modified version of the system of KLINKA et al. (1981) was used to describe the organic horizons, which in that system are defined as the ectorganic horizons. KLINKA et al. distinguish L (litter), F (fermentation) and H (humus horizons, and subdivide the F horizon into Fa, Fq and Faq horizons, depending on the predominance of faunal (a) or fungal (q) activities. Transitions between L and F horizons and between F and H horizons, however, generally appeared to be very gradual and therefore F1, F2 and F3 horizons were distinguished. The F1 horizon is transitional to the L horizon, the F2 is a true fermentation horizon and the F3 is transitional to the H horizon.

Chemical anayses were carried out on 1:10 water extracts of the soil samples, after filtration of the extracts over a 0.45

Site	L mean	L s.d.	F1 mean	F1 s.d.	F2 mean	F2 s.d.	F3 mean	F3 s.d.	H mean	H s.d.
5	.76	.45	.28	.37	.61	.75	.23	.39	.32	.96
4	.99	.61	.29	.65	.71	1.13	.75	.45	.29	.71
11	1.06	.83	.37	.53	.78	.95	1.64	1.95	.01	.08
12	.89	.47	.47	.72	.96	.77	1.27	1.43	.02	.06
14	.44	.22	.82	.63	.28	.28	.79	1.16	-	-
13	1.87	1.60	.82	.87	.89	.64	.69	.90	-	.02
1	1.83	1.74	.88	1.52	.73	.94	1.33	1.56	-	-
8	1.58	1.28	.95	1.15	2.03	1.64	2.44	2.58	.19	.58
7	1.41	1.04	1.87	1.71	1.98	1.21	2.60	3.05	.64	1.43
6	1.02	.91	.90	1.26	1.60	2.39	1.28	1.87	-	.03
9	1.15	.90	.25	.44	.86	.87	.60	1.18	.05	.16

Tab. 3: *Thicknesses of individual ectorganic horizons: mean thickness (m) and standard deviations (S.D.), both in cm.*

µm membrane filter. N species, P species, Cl and S were estimated by Auto Analyzer, while K, Na, Ca and Mg were estimated by Atomic Absorption Spectrophotometer. Soil reaction was determined in a suspension of 10 gr soil in 25 ml water and in 25 ml of a 0.01 M CaC12 solution.

The subsequent research, carried out in summer 1986, was more specific and concentrated on plots with a distinctly mediterranean climate on granodiorite or on sediments, derived from these rocks. Aspects studied were the effects of species composition of the vegetation, forest management practices, forest fires and erosion on the characteristics of and the spatial variability in organic matter profiles. To that purpose systematic detailed surveys of organic matter profile characteristics and of the vegetation were carried out on a series of plots near Santa Coloma de Farnes and Santa Cristina d'Aro. Plot sizes were 10 × 10 m and the surveys were at scale 1:50.

With the aid of a rainfall simulator, the hydrological characteristics of representative soils within the plots were studied. The experiments were carried out, using intensities of about 40 mm/hr, applied for 30 minutes, on an area of 0.5 m^2.

4 The General Trends

In tab.3 the various organic horizons encountered in the plots, their mean thickness and the standard deviation in their thickness are indicated, while part of the data are graphically presented in fig.2. The first conclusion which can be drawn is that the data confirm the earlier research in Italy, which indicated that under mediteranean climatic conditions accumulation of organic matter occurs and that turnover rates of litter are certainly not as high as could be expected in view of the current pedogenetic theories, which assume that a mull type humus is formed. The data furthermore show that the spatial variability on the whole is strong, i.e. at micro-scale (less than 2 m) large variations in horizon succession and horizon thickness may occur.

The data indicate that at lower altitudes parent material is a major factor controlling the organic profile development and that exposition plays a

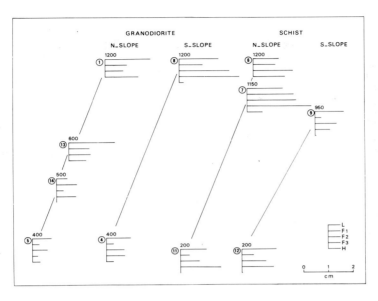

Fig. 2: *Graphic representation of mean thickness of individual ectorganic horizons.*

Sample	pH H$_2$O	pH CaCl$_2$	K$^+$	Na$^+$	Ca^{2+}	Mg^{2+}	N$_{tot}$	% NH$_4^+$	% N org.	% NO$_3^-$	Cl^l	SO$_4^{2-}$
4.L	4.77	4.17	4095	318	1952	2041	3189	9	90	1.3	237	1860
4.F	5.42	4.87	1777	298	1890	836	4058	17	83	.4	212	691
5.L	4.60	4.29	4322	344	2000	1927	2717	6	93	1.2	521	2810
5.F	4.84	4.36	2420	289	2096	1306	4417	11	89	.5	291	1175
12.L	4.37	4.08	4105	347	3849	3562	2075	8	90	2.0	409	5500
12.F	4.65	4.19	1050	198	1732	1118	2853	14	85	.5	216	871
11.L	4.41	4.12	6131	364	3870	2924	2650	7	92	2.0	415	4930
11.F	5.06	4.55	1307	214	1597	900	2721	19	80	1.0	243	467
14.L	4.33	4.01	9360	443	2141	4191	3623	7	86	7.0	490	8030
14.F1	4.80	4.42	4475	723	3341	3080	7013	10	89	1.2	483	3430
14.F2	5.72	5.33	4885	1115	3918	3867	9469	42	57	.6	1113	994
14.F3	6.88	6.25	2960	605	2381	2111	5185	37	44	19.0	560	124
13.L	4.93	4.42	1522	378	1245	3389	1542	19	80	1.4	362	589
13.F1	5.62	5.09	587	246	1463	1040	1833	28	71	1.3	170	102
13.F3	5.47	5.04	648	209	1595	953	1810	32	67	.6	160	102
1.L	5.07	4.70	1277	220	1595	774	3430	23	77	.7	172	263
1.F1	5.38	4.81	1534	352	1877	1008	4330	20	80	.6	277	348
1.F2	5.88	5.29	1429	253	2271	852	5663	49	51	.2	280	274
1.F3	6.01	5.50	373	250	1151	387	3425	26	38	36.0	194	-
7.L	5.67	5.03	1477	320	1835	874	3471	23	76	.5	233	226
7.F1	5.58	5.04	1384	227	2168	1042	4308	20	80	.3	207	270
7.F2	6.94	6.00	1112	635	2237	617	6881	54	42	3.7	573	249
7.F3	6.80	5.76	1176	404	1465	416	4982	60	36	3.6	-	360

Tab. 4: *Chemical composition of 1:10 water extracts of ectorganic horizons.*

very subordinate role. That vegetation also strongly controls its development is demonstrated by the data from the plots 13 and 14, under Pinus and Castanea plantations respectively.

At higher altitudes relations become more complex, although the general trend is that the accumulation of organic matter is more prominent. Effects of differences in vegetation are reflected in plots 6 and 7, the soils under Fagus forest showing a stronger accumulation of organic matter than under an open Fagus-Juniperus vegetation. Comparison of plots 1 and 8, and of plots 1 and 6 suggests that at that altitude (1200 m) litter decomposes relatively rapidly on the south facing slopes and in soils on granodiorite.

In tab.4 data on the chemical composition of 1:10 water extracts from soils within the mediteranean climatic zone are presented. For comparison also data on two soils under beech forest (plots 1 and 7) are presented.

The major characteristics of the plots with distinctly mediterranean vegetation (4, 5, 11 and 12) are:

- the relatively high concentrations of K, Ca and Mg,

- the relatively high concentrations of Nt, mainly in the form of organic N, pointing to a low degree of mineralization,

- the high concentrations of SO4.

Differences in exposition apparently do not lead to marked differences in chemistry, but parent material seems to play an important role: on schist L horizons are relatively high in Ca, Mg and SO4, whereas F horizons are relatively low in K, Nt and SO4.

The differences between the "mediterranean" and "beech" plots are clear and concern in particular pH, which is higher under beech, and K, Mg and SO4, which are distinctly lower. Furthermore, under beech nitrogen mineralization is stronger.

The Castanea plot 14 shows by far the highest element concentrations of all plots and a distribution pattern and composition which completely deviates from that in the "mediterranean" plots. The Pinus plot, on the contrary, shows relatively low concentrations.

Combining the data discussed above, some general statements can be made on the litter decomposition in the attenuated mediterranean climatic zone.

For the profiles under a sclerophyllous vegetation (*Quercus suber, Arbutus unedo, Erica arborea, Pinus halepensis*) the accumulation of organic matter, presence of F and H horizons and poor mineralization of organic N point to a rather slow and incomplete decomposition of the litter, rather irrespective of the species composition of the vegetation and of the parent material. The high contents of water extractable elements, however, point to a rapid mineralization of litter in the upper organic horizons, while the observed changes with depth in K, Mg and SO4 indicate that nutrient uptake in the lower organic horizons is an important process. Evidently the nutrient recycling is rather strong and, in so far as the soil is concerned, largely takes place in the organic horizons. It should be stated that the latter conclusion is in accordance with the observations on the root distribution pattern, the fine roots being largely concentrated in the F and H horizons.

The observed differences in chemistry between the plots on schist and on granodiorite can be easily explained as being due to differences in composition of the parent material and indicate that the latter differences influence the elemental composition of the litter, rather than total levels of concentration.

The seemingly conflicting observations on the litter decomposition might be explained by assuming that during a first phase of intensive but partial litter breakdown large amounts of elements are released, followed by a sharp decrease in turnover rate of the remaining litter material. Indications for such a sequence of processes can be found in the literature (READ & MITCHELL 1983).

The data on the soils under *Castanea sativa* forest point to a completely different type of decomposition: a rapid turnover of all litter and concurrent massive release of elements contained in these in the topsoil. The comparatively high levels of water soluble elements are most likely to be attributed to the relatively high litter input under these deciduous trees. Under Pine forest, on the contrary, the litter turnover rate seems to be low, taking also into account the presumably large difference in litter input between these two types of vegetation. Considering the concentrations observed, this rate might be the lowest of all mediterranean plots studied. As the differences in altitude between the plots 5, 13 and 14 are too small to cause major differences in climate, it must be concluded that the observed differences are largely due to differences in composition and quantity of the litter.

5 The Specific Research

The effects of individual plant species on the organic matter profile characteristics were clearly traceable, being reflected mainly in the structure of the various ectorganic horizons and in the presence or absence of an H horizon. It could also be clearly inferred that transport of litter by surface runoff is an important process and one of the factors inducing a strong spatial variability in thickness of the ectorganic horizons. This is illustrated by maps of two of the plots, presented in the figs.3 and 4.

Fig.3 shows the spatial variations in litter deposition and composition in a dense stand of *Quercus suber* and *Arbutus unedo* on a flat summit. Fig.4 depicts the various types of organic matter profiles in a mixed forest stand on a ridge. It shows that changes in thickness of the profiles are rather systematically associated with slope breaks, pointing to litter transport by surface runoff.

The surveys of plots with recently burnt vegetation indicate that the effect of a fire on the ectorganic horizons depends on the intensity of the fire. High intensity fires cause the complete destruction of all such horizons, whereas after low intensity fires (i.e. most trees and shrubs survived the fire) the H horizon is still preserved. It should be stressed, however, that this concerns the overall pattern, but that at microscale large variations in the effect of fire can be observed.

Burnt plots show prominent effects of hill slope runoff, such as locally massive colluvial accumulations of more or less finely divided charcoal and of coarse sandy to fine gravelly rock fragments (mainly quartz), often covering preserved H horizons.

Intensive cutting of undergrowth, as commonly practiced in cork oak stands, was found to lead to a rapid disappearance of the L and F horizons, while H horizons are generally preserved. Hill

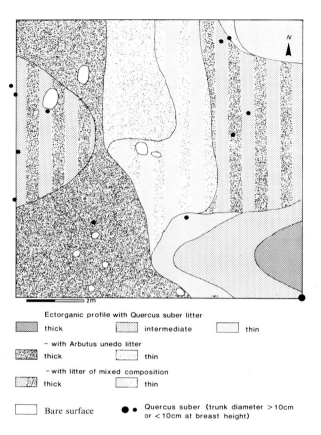

Fig. 3: *Detailed map showing species dependent variations in litter composition and input under a Quercus suber and Arbutus unedo stand on a level summit near Santa Cristina d'Aro. For tree symbols, see fig.4.*

slope runoff seems to be largely responsible for this phenomenon, as evidenced by the presence of colluvium, identical to that observed on burnt plots, but for the charcoal which is absent.

Because of the preliminary character of the rainfall simulator experiments, results can only be presented in rather descriptive terms.

Ectorganic horizons (L, F and H) of plots under (semi)natural vegetation have a relatively high infiltration and storage capacity: ponding on these horizons was not observed and during the experiments, depending on the thickness of the ectorganic horizons, between 10 and 15 mm of rain was required before the wetting front reached the mineral horizon. In contrast, the mineral horizon immediately underneath (Ah horizon, generally of limited thickness) is strongly water repellent, starting to accept water only one hour or more after the end of the experiment. This prevented deeper infiltration after saturation of the ectorganic horizons and caused throughflow over the mineral soil, once the ectorganic horizons were saturated.

Incidentally, large vertical macropores occur, which are partially filled with non-water repellent material and conduct water to deeper horizons. This leads to a very irregular wetting front and penetration of water, resulting in large disconti-

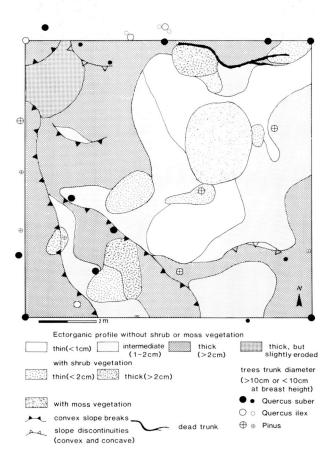

Fig. 4: *Detailed map of the ectorganic profiles and slope breaks on a sloping ridge near Santa Cristina d'Aro.*

nuities in water content within the soil. These pores most probably are biopores, formed by larger soil fauna, including earthworms. The number of large pores per unit surface area, althoug small, is quite variable and seems to control the ratio between the throughflow and deep infiltration. Where present they were found to prevent throughflow even at rainfall intensities of 60 cm per hour, maintained for 30 minutes.

On burnt plots a patchy pattern in water repellency was observed. Where parts of the ectorganic profile are preserved (mainly H horizons, see above) the mineral horizon below is generally water repellent (no acceptance of water within 30 minutes) and throughflow over the mineral soil starts after saturation of the relict ectorganic horizons. In the absence of ectorganic horizons, water repellency is weak (acceptance within several minutes) or absent. However, the infiltration capacity of the upper mineral horizon is insufficient for the rainfall intensity used (40mm/hour) and overlandflow rapidly starts (i.e. after a few minutes).

In burnt plots soils with and without ectorganic horizons occur in an irregular, intricate pattern. The distinction between throughflow and overlandflow in these plots is rather theoretical, all water flowing over the mineral soil. Experiments on a number of burnt plots showed that the magnitude of this flow depends strongly on the extent to which ectorganic horizons are preserved because of their relatively high storage capacity. Quantitative studies on this relation have not yet been carried out.

The soils under intensively managed *Quercus suber* forest have characteristics, strongly resembling those under (semi)natural vegetation: the upper mineral horizon is also strongly water repellent (see above) and similar large biopores occur. However, with decreasing thickness of the ectorganic horizons the time required to saturate these horizons decreases and throughflow starts more rapidly. Furthermore, because of the lower storage capacity of these horizons the throughflow is more intense. Where only H horizons are preserved, the phenomena observed were in fact similar to those of plots affected by low intensity fires, i.e. throughflow started rapidly (within about 10 minutes).

6 Conclusions

The preliminary conclusions of the research, presented here, deal with the summer situation only. Further research into the seasonal and spatial variations in soil characteristics is required, as well as a more quantitative study of the processes active and of the infiltration characteristics and storage capacities of the ectorganic horizons. However, the present results already clearly indicate that large differences can be expected in runoff from soils on slopes in acid rocks, in particular in its quantity and chemistry, and that these differences are strongly determined by forest management practices and forest fires. The results can be summarized as follows:

- In contrast to what is generally assumed, accumulation of organic matter due to a relatively slow decomposition of the litter is an important process in mediterranean forests and strongly determines the physical and chemical characteristics of the topsoil.

- Mediterranean forest soils, hardly affected by man or fire, are not likely to generate hillslope runoff. Hwever, during the dry season, after rainstorms throughflow over the mineral soil may occur. High concentrations of N, K, Ca and Mg are to be found in the runoff, but transport of mineral material will be minimal due to the protecting action of the ectorganic horizons and the roots contained in these.

- Upon forest fires, due to the destruction of ectorganic horizons and of the roots in these horizons, as well as of the protective vegetation cover, overlandflow will readily occur after rainstorms and will easily lead to accelerated soil erosion. Initially, solute concentrations will be high, but after leaching of the remaining ashes these will be rapidly lowered and probably reach values far below those observed under (semi)natural vegetation. However, in case that fires are less intense and parts of the ectorganic horizons are preserved, the effects of fire are probably less intense.

- Intensive cutting of the understorey in (semi)natural forests, as practiced in cork oak forests, seems to lead to a significant reduction of the L and F horizons and to an increase in hillslope runoff and erosion. Here too, solute concentrations initially will be high and will rapidly decrease, but changes in solute concentrations will probably not be as prominent as after fires.

Acknowledgement

Thanks are due to the postgraduate students R. Bol, M. Boots, I. Emmer, N. Sinnige, N.H. Stikker and P. Veerkamp, who carried out most of the laborious inventory research, and to Prof. J.M. Verstraten and Dr. A. Imeson, who both very actively participated in the specific research and in the scientific discussions.

References

BOLOS, DE O. (1983): La Vegetacio del Montseny; Diputacio de Barcelona. Servei de Parcs Naturals, Barcelona.

BREMEN, VAN H. (1985): Het ontstaan van bodems op low-grade metamorfe gesteenten van de Verrucano formatie in de Monte Leoni (Toscane, Italie). Internal report Laboratory for Physical Geography and Soil Science, University of Amsterdam.

BREMEN, VAN H. (1986): Humus profiles in Mediterranean and Sub-mediterranean Ecosystems: Morphological and chemical diversity in the Farma Valley (Italy). Internal report Laboratory for Physical Geography and Soil Science, University of Amsterdam.

CASTRI, DI F., MOONEY, H.A. (eds.) (1973): Mediterranean type ecosystems: Origin and structure. Springer Verlag, Hamburg / Berlin.

DUCHAUFOUR, P. (1982): Pedology. George, Allen and Unwin, London.

FAO / UNESCO (1977): Guidelines for soil profile description. Land and Water Development Division, FAO, Rome.

KLINKA, K., GREEN, R.N., TROWBRIDGE, R.L. & LOWE, L.E. (1981): Taxonomic Classification of Humus Forms in Ecosystems of British Columbia. First Approximation. Ministry of Forests, British Columbia.

KRUGER, F.J., MITCHELL, D.T. & JARVIS, J.U.M. (1983): Mediterranean type ecosystems: The role of nutrients. Springer Verlag, Hamburg/ Berlin.

PEDROLI, G.B.M., SEVINK, J. & VOS, W. (1985): Landscape-ecological classification and valuation: The Farma case. Transactions meeting IGU working group on "Landscape Synthesis", Dessau.

READ, D.J. & MITCHELL, D.T. (1983): Decomposition and Mineralization Processes in Mediterranean-Type Ecosystems and in Heathlands of Similar Structure. In: Kruger, F.J., Mitchell, D.T. & Jarvis, J.U.M.: Mediterranean type ecosystems: The role of nutrients. 208–232, Springer Verlag, Hamburg/Berlin.

SALA, M. (1979): L'organitzacio de l'espai natural a les Gavarres. Dalmau, Barcelona.

VOS, W. & PEDROLI, B. (Eds.) (1985): The Farma Barrage Effect Study; a landscape impact procedure applied to the Farma drainage basin water derivation plans (Tuscany, Italy). 2 reports Regione Toscana, Firenze, Italy.

Address of author:
J. Sevink
Laboratory of Physical Geography and Soil Science
University of Amsterdam
Dapperstraat 15
1093 BS Amsterdam, The Netherlands

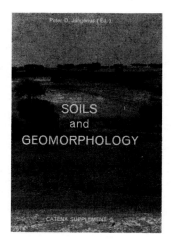

Peter D. Jungerius (Ed.):

Soils and Geomorphology

CATENA SUPPLEMENT 6 (1985)

Price DM 120,-

ISSN 0722-0723 / ISBN 3-923381-05-0

It was 12 years ago that CATENA's first issue was published with its ambitious subtitle "Interdisciplinary Journal of Geomorphology – Hydrology – Pedology". Out of the nearly one hundred papers that have been published in the regular issues since then, one-third have been concerned with subjects of a combined geomorphological and pedological nature. Last year it was decided to devote SUPPLEMENT 6 to the integration of these two disciplines. Apart from assembling a number of papers which are representative of the integrated approach, I have taken the opportunity to evaluate the character of the integration in an introductory paper. I have not attempted to cover the whole bibliography on the subject: an on-line consultation of the Georef files carried out on 29th October, 1984, produced 3627 titles under the combined keywords 'geomorphology' and 'soils'. Rather, I have made use of the ample material published in CATENA to emphasize certain points.

In spite of the fact that land forms as well as soils are largely formed by the same environmental factors, geomorphology and pedology have different roots and have developed along different lines. Papers which truly emanate the two lines of thinking are therefore relatively rare. This is regrettable because grafting the methodology of the one discipline onto research topics of the other often adds a new dimension to the framework in which the research is carried out. It is the aim of this SUPPLEMENT to stimulate the cross-fertilization of the two disciplines.

The papers are grouped into 5 categories: 1) the response of soil to erosion processes, 2) soils and slope development, 3) soils and land forms, 4) the age of soils and land forms, and 5) weathering (including karst).

P.D. Jungerius

P.D. JUNGERIUS
  SOILS AND GEOMORPHOLOGY

The response of soil to erosion processes
C.H. QUANSAH
  THE EFFECT OF SOIL TYPE, SLOPE, FLOWRATE AND THEIR INTERACTIONS ON DETACHMENT BY OVERLAND FLOW WITH AND WITHOUT RAIN
D.L. JOHNSON
  SOIL THICKNESS PROCESSES

Soils and slope development
M. WIEDER, A. YAIR & A. ARZI
  CATENARY SOIL RELATIONSHIPS ON ARID HILLSLOPES
D.C. MARRON
  COLLUVIUM IN BEDROCK HOLLOWS ON STEEP SLOPES, REDWOOD CREEK DRAINAGE BASIN, NORTHWESTERN CALIFORNIA

Soil and landforms
D.J. BRIGGS & E.K. SHISHIRA
  SOIL VARIABILITY IN GEOMORPHOLOGICALLY DEFINED SURVEY UNITS IN THE ALBUDEITE AREA OF MURCIA PROVINCE, SPAIN

C.B. CRAMPTON
  COMPACTED SOIL HORIZONS IN WESTERN CANADA

The age of soils and landforms
D.C. VAN DIJK
  SOIL GEOMORPHIC HISTORY OF THE TARA CLAY PLAINS S.E. QUEENSLAND
H. WIECHMANN & H. ZEPP
  ZUR MORPHOGENETISCHEN BEDEUTUNG DER GRAULEHME IN DER NORDEIFEL
M.J. GUCCIONE
  QUANTITATIVE ESTIMATES OF CLAY-MINERAL ALTERATION IN A SOIL CHRONOSEQUENCE IN MISSOURI, U.S.A.

Weathering (including Karst)
A.W. MANN & C.D. OLLIER
  CHEMICAL DIFFUSION AND FERRICRETE FORMATION
M. GAIFFE & S. BRUCKERT
  ANALYSE DES TRANSPORTS DE MATIERES ET DES PROCESSUS PEDOGENETIQUES IMPLIQUES DANS LES CHAINES DE SOLS DU KARST JURASSIEN

SOIL DEVELOPMENT AND GEOMORPHIC PROCESSES IN A CHAPARRAL WATERSHED: RATTLESNAKE CANYON, S. CALIFORNIA, USA

A.G. **Brown**, Leicester

Summary

An investigation of soil development, in a steep chaparral basin employing magnetic susceptibility measurements demonstrates considerable spatial variations due to lithology and slope history. All the soils are magnetically enhanced due to the Mediterranean climate and fires, but, the enhancement varies with lithology and geomorphic activity such as landsliding. Contrasts between laboratory induced magnetic susceptibility enhancement and field enhancement can be related to soil erosion and soil development ranging from dry-ravel erosion on shales to pocket soil development on sandstones.

Resumen

La investigación sobre el desarrollo de los suelos de una cuenca con fuertes pendientes y vegetación de tipo chaparral, llevada a cabo mediante mediciones de la susceptibilidad magnética, muestra la existencia de considerables variaciones espaciales relacionadas con la evolución geomórfica de la litologia y las vertientes. Todos los suelos presentan un magnetismo intensificado debido al clima mediterráneo y a los incendios, pero esta intensidad varía tanto en función de la litología como de procesos geomórficos tales como los deslizamientos. Los contrastes entre el aumento inducido de la susceptibilidad magnética en el laboratorio y el aumento en el campo pueden ser relacionados con los contrastes entre àreas con fuerte erosión, como en el caso de los sectores de pizarras, y áreas con un buen desarrollo de bolsas de suelo, como en el caso de los sectores de areniscas.

1 Aims and Introduction

The development of soils, and the rate of downslope transport of soil and weathered rock, are an important part of the total sediment budget of any basin. The depth and characteristics of soils are related to the rates of bedrock weathering, aerial input and loss of material in solution and by soil erosion and mass movement. If the rates of input exceed the rates of output soils will thicken and develop. There is therefore a positive relationship between soil development and age given a degree of geomorphic stability and progressive pedogenesis (JOHN-

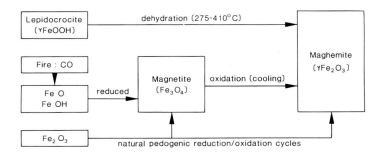

Fig. 1: *Summary diagram of soil processes responsible for magnetic susceptibility enhancement.*

SON 1985).

The aim of the work described here is to investigate variations in soil development on the different lithologies in relation to the processes and rates of erosion, and to evaluate the extent to which soil development is related to soil stability and inversely related to soil erosion in a chaparral watershed. The principal method used was magnetic susceptibility (χ) because the enhancemment of magnetic susceptibility in soils is known to be related to age and geomorphic stability as well as other pedological factors.

1.1 Soil Magnetic Susceptibility

The common magnetic enhancement of topsoil discovered by LA BORGNE (1955) is generally due to magnetite (Fe_3O_4) and/or maghemite ($\gamma\ Fe_2O_3$). Due to similar crystal structures these two minerals are difficult to differentiate using traditional techniques such as XRD (MULLINS 1977). In soils not containing large amounts of inherited magnetite high susceptibility may be due to the presence of maghemite which can be formed by four processes. These processes are: low temperature oxidation of magnetite; burning; dehydration of lepidocrocite ($\gamma\ FeOOH$) and reduction-oxidation cycles which can occur under normal pedogenic conditions (fig.1). For a more detailed description of these processes see MULLINS' (1977) review of magnetic susceptibility and soils.

The magnetic enhancement is generally carried by the clay sized fraction and will reflect not only the initial content of iron (which can by converted into ferrimagnetic forms) in the underlying rocks but the length of time the soil has been subjected to oxidation-reduction cycles without significant topsoil erosion. As fire has been shown to be very effective in increasing the magnetic susceptibility of soils (TITE & MULLINS 1971) soils under Mediterranean climates with Chaparral vegetation might be expected to display high enhancement. TITE & LININGTON (1975) have shown that soils under Mediterranean climates do exhibit high magnetic susceptibility enhancement. It has also been shown that maghemite formation is favoured by high temperatures and high pH (TAYLOR & SCHWERTMANN 1974). However, due to the complexities of the processes responsible for the magnetic enhancement of soils, there will not be a linear relationship between soil age and magnetic susceptibility, nor may magnetic susceptibility be used to measure the quantity of ferrimagnetic iron in any soil. But on rocks with relatively low background magnetic susceptibility it is possible to use the magnetic properties of soils to investigate relative rates of soil stability and erosion, since both naturally occurring pedogenic oxidation-reduction cycles and fires are time dependent given soils of a similar type and under similar climatic conditions. Due to the complexity of soil conditions responsible for the production of maghemite, magnetic susceptibility may not be used as a general indicator of soil forming processes (MULLINS 1977). It will, however, display an inverse relationship with soil erosion and therefore may be an indicator of geomorphic processes (DEARING et al. 1985).

2 Rattlesnake Canyon

Rattlesnake Canyon lies at the edge of the Los Padres National Forest to the north of Santa Barbara in Southern California. It is a steep coastal watershed with a relief of 900 m and a basin area of 8.2 km^2 (fig.2). The climate of the region is Mediterranean with episodic rainfall between November and March and a summer drought. Steep chaparral basins, such as Rattlesnake Canyon are characterised by high erosion rates especially after fires (DE BANO et al. 1979), which cause soil shallowing and regressive pedogenesis. Transport of sediment within and from the basin by large but infrequent floods along with fire disturbance produces what RICE et al. (1892) call hyperscedastic basin response. The detailed geology and hydrometeorology can be found in KELLER et al. (1985). The topography within the basin, and inverted relief, suggest quite different erosion rates on the different lithologies in an area with a rapid uplift rate (KELLER et al. 1985). The geology is southwesterly dipping shales and sandstones, frequently exposed on dip slopes and cliffs on strike valleys. Soils are generally thin and poorly developed. The soils are classified as Mayman-rock outcrop complex (MbH) on 50–100% slopes at the top and bottom of the basin, and Rock Outcrop Mayman complex (Rb) in the middle of the basin. Both of these soil complexes have a very high erosion potential. The MbH complex has an available water capacity of only 2.5–5.0 cm and an effective rooting depth 25–46 cm and is typical on the shale. The Rb complex has soils that are generally over 38 cm deep and have an available water capacity of 0–3.8 cm with a low permeability of 1.5–5.0 cm hr^{-1}. Storm intensities can reach 15 cm hr^{-1} (TAYLOR 1983). Both soils will produce rapid and excessive runoff and have a high erosion hazard. The vegetation of Rattlesnake Canyon is mostly hard Chaparral dominated by *Ceonothus* spp., *Adenostoma fasciculatum* and *Arctostaphylos* spp. Like savanna grassland, eucalyptus and ponderosa pine forest, chaparral vegetation is ecologically adapted to frequent fires (MUTCH 1970). Soils in the basin are generally circum-neutral to slightly acidic (see tab.2) and vary from around pH 5.4 on the Matilija sandstone to pH 6.8 on the Juncal shale. The % carbon (% C in tab.2) content of the under 2 mm fraction in relation to the loss on ignition values (% LOI) illustrates how much of the organic matter is undecomposed woody litter which is incorporated into the A horizon. The high levels of organic matter do not indicate the development of organic rich A horizons but the rapid production of resistant litter by chaparral vegetation

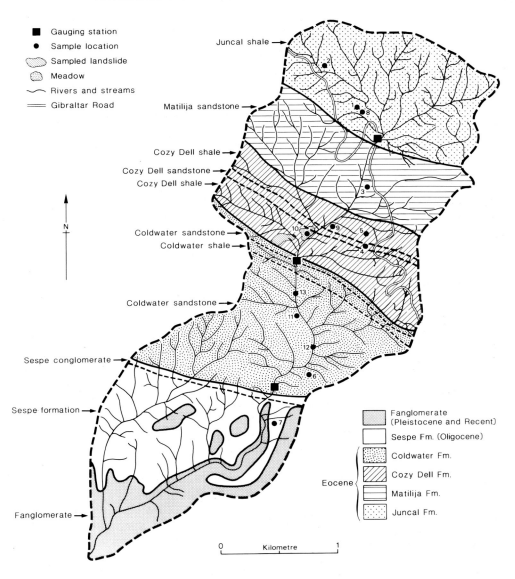

Fig. 2: *Map of Rattlesnake Canyon showing geology and main sampling sites.*

and a lack of decomposition.

Photo 1: *Juncal soil site. Tape marked in cm, pen top is 2.5 cm tall.*

Photo 2: *Matilija sandstone soil. Tape marked in cm.*

There is, however, considerable variation within these soil complexes due to relief and lithology. Soils on the Juncal formation, which is a thin-bedded, fissile mudstone, are commonly very thin and lack any B horizon and many slopes are covered only by a thin (ca. 8–10 cm) dry-ravel layer or a gravelly sandy loam. The soils are composed of silt with angular fragments of shale and some completely lack any soil structure being single grain apedal in character (photo 1). In contrast soils of the Matilija formation are very variable but can be much deeper and generally include a thin sandy B horizon with some silt, although they too can be apedal (photo 2). This horizon is non-wettable but the generally thick litter layer of these soils allows relatively high infiltration rates, reducing the erosion risk. The Cozy Dell formation contains both sandstones and shales. Soils on the sandstones are similar to those on the Matilija formation, but soils on the shales are thicker and better developed than those on the Juncal shale with a weak to moderate blocky structure. Soils on the sandstones and shales of the Coldwater formation are relatively well developed with blocky structure and thin sandy B horizons while soils on the sandstones and conglomerate of the Sespe formation are red and the deepest in the watershed with strongly developed structure and occasional cemented B horizons.

Pits were dug both in these slope soils and also in a number of sediment sinks, such as valley bottoms and slope bases. These are generally cumulative soils. On the Juncal formation 2.2 m of alluvial and colluvial deposits was exposed by stream incision. A basal alluvial unit of closely packed gravel and cobbles was found to underlie a silt unit with angular fragments of Juncal shale, interpreted as being of colluvial origin. Between 180 and 200 cm was a caliche layer.

Formation	Lithology	Bedrock magnetic susceptibility ($\times 10^{-6}$ G.Oe^{-1}cm^{-3}g^{-1})	Mean soil magnetic enhancement (%)	Soil depth (cm)	Resistance to erosion (DIBBLEE 1966)
Juncal	grey clay shale	13	723	9	very low-moderate
Matilija	buff sandstone	2	1,400	28	very high
Cozy Dell	grey clay shale	10	3.120	11	very low
	buff sandstone	2	15,850	22	high
Coldwater	siltstone	3	1.366	80+	moderate but locally
	shelly sandstone	5	820	80+	low to very high
Sespe	pink arkosic sandstone	8	1.650	90	moderate
	red conglomerate	8	1.650	130	very high
Alluvial fan (on coldwater sandstone)	boulder/gravel gravel, sand and silt	43	325	5	very low

Tab. 1: *Unweathered bedrock magnetic susceptibility, soil magnetic enhancement, soil depth and resistance to erosion.*

On the Cozy Dell formation two pits showed deep cumulative soils with overthickened black/brown A horizons, one overlying alluvial gravel the other with an unknown deposit beneath it. On the Coldwater formation the present stream is cutting into an alluvial/debris fan deposit of over 4 m thickness. At the base is a closely packed cobble gravel with a sand and gravel unit above it with matrix supported cobbles and occasional boulders, some of which lie on the surface. Below 270 cm the sediments are cemented by calcium carbonate. A mid-channel bar in the Coldwater sandstone reach displays very little soil development with only a thin A horizon lying directly over sand and gravel which coarsens with depth. Lastly sediments were sampled from a tunnel in the stream channel which was probably built in the nineteenth century for water supply. The black organic sand and silt infill must have been derived entirely from stream transport. Most of the sediment stores are at the present time being eroded and are thus sediment contributing sites. Other sites of sediment contribution are the stream bed and rocks which fall from free faces above the channel in the sandstone reaches and weather **in situ**.

3 Methods and Results

Large bulk samples taken from the full thickness of each soil horizon were homogenised and subsampled. From the coarse units of the sediment sites small matrix samples were taken at regular intervals, homogenised and sub-sampled. The magnetic susceptibility of the soils and sediments was measured using a Bartington magnetic susceptibility meter generating a magnetic field of 1 Oe. and measurements are expressed in G.Oe.$^{-1}$cm^{-3}g^{-1} ($\times 10^{-6}$) mass susceptibility.

As shown in tab.1 the magnetic susceptibility of the unweathered rocks within the catchment is low, with the shales having slightly higher values than the sandstones. The Coldwater and Matilija sandstones are both very pure quartzoze marine sandstones while the Sespe is terrestrical and contains pebbly arkosic sandstones with some haematite as well as shale units. The Juncal and Cozy Dell shales both display iron oxide films on fracture surfaces. As none of the rocks in the watershed have a high magnetic susceptibility this allows enhancement to be easily picked up.

3.1 Results: Soils

All the slope soils within the catchment show enhancement of the A horizons (fig.3) some to a very high degree with enhancement varying from ×2 to nearly ×160 (tab.1). This enhancement is high in comparison with other figures from temperate areas (LE BORGNE 1955, TITE & LININGTON 1975) but similar to those from freely draining Mediteranean soil (TITE & LININGTON 1975). This is probably due to the combined effects of good reducing conditions during the mild winter, oxidising conditions during the summer and the frequency of fires in these regions. In chaparral watersheds fires have a return period of approximately 50 years and the last fire in Rattlesnake Canyon occurred in 1964. However, enhancement varies with rock type with the thin Juncal soils exhibiting the lowest enhancement despite the highest rock susceptibility which suggests a lack of soil stability (and therefore "young" soil) due to high erosion rates. In contrast the high enhancement of the Cozy Dell sandstone reflects greater soil stability, better profile development and lower erosion rates. The Cozy Dell shale soils are also better developed than those on the Juncal and have a much higher magnetic enhancement. The Matilija and Coldwater sandstones have medium enhancement while the Sespe is higher probably because of its initially higher (anti-ferrimagnetic) iron content and the relative stability of the soils, depth and profile development. The Sespe soil is the ony one to show maximum enhancement in the A_2 horizon probably due to the translocation of clay sized material. Where possible samples were fractionated by dry sieving and the magnetic susceptibility of the under 53 μm fraction (silt and clay) was measured (tab.2). In nearly all cases the figures indicate that it is this fraction or smaller that is carrying the magnetic enhancement.

Fires effecting the basin are not restricted to any particular rock types and the vegetational pattern tends to reflect topography, altitude, aspect and geology.

3.2 Landslide Transect

One of the principal processes of erosion in the watershed is mass-movements and especially landslides. They are most frequent on the steep scarp slopes of the strike valleys. To investigate the effect of a landslide on soil magnetic properties a transect was run across the largest landslide in the watershed. The landslide scar is approximately 400 m long and 30–200 m across. At its top is a free face formed by the junction of the Coldwater shale and the Coldwater sandstone. The sides are steep reaching 55° and down its axis runs a stepped bedrock channel at a 30° slope. Aerial photography revealed that the landslide occurred before 1929. Fig.4 shows how the magnetic susceptibility of the landslide soils is much lower than that of the soils around it. It is suggested that this is due to the thin, poorly developed soils on the land-

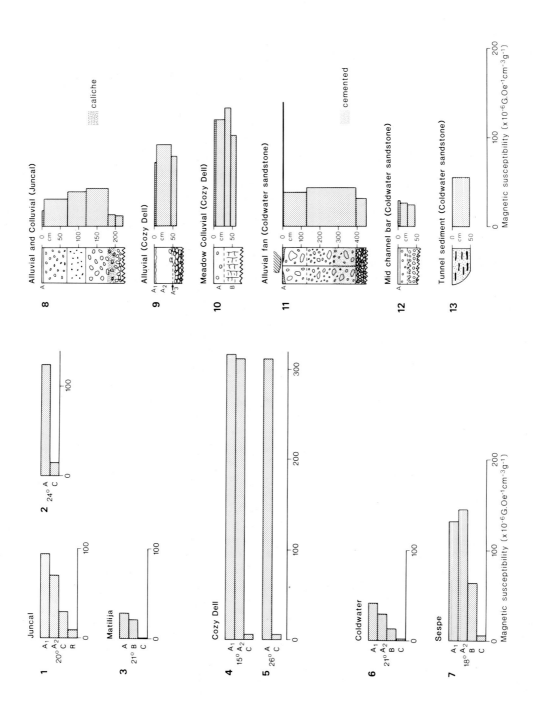

Fig. 3: *Magnetic susceptibility profiles from main sites. Numbers refer to fig.2 and slope of the site is indicated.*

Soil Development in a Chaparral Watershed

Soil location (A horizons)	pH	LOI (%wt)	C (%wt)	Fe (%wt)	χ field	Magnetic susceptibility χ field (< 53μm) x 10^{-6} G. $Oe^{-1}cm^{-3}g^{-1}$	χ air 550°C	χ N 550°
Juncal shale	6.8	16	9.2	2.0	94	182	3316	1841
Matilija sandstone	5.4	25	6.8	2.0	28	-	784	845
Cozy Dell	5.7	12	3.4	1.6	317	240	3007	1201
Coldwater sandstone	6.4	6	1.7	1.1	41	-	675	583
Sespe sandstone	6.0	5	1.7	1.1	132	221	953	851
Alluvial/Colluvial on Juncal	6.3	8	1.9	1.7	18	-	-	-
Alluvial on Cozy Dell	6.7	8	2.9	1.1	72	-	-	-
Colluvial (meadow)	6.4	10	3.0	1.7	120	126	-	-
Alluvial fan on Coldwater	5.6	6	1.4	1.3	140	-	-	-
Mid-channel Bar	6.7	19	5.7	0.7	29	-	-	-
Landslide Edge	6.5	27	6.7	0.7	212	-	-	-
Landslide Middle	6.6	6	2.3	1.2	30	-	-	-
Burnt soil Ojay	-	-	-	-	696	621	783	-

Tab. 2: *Soil chemical and magnetic data from key horizons.*

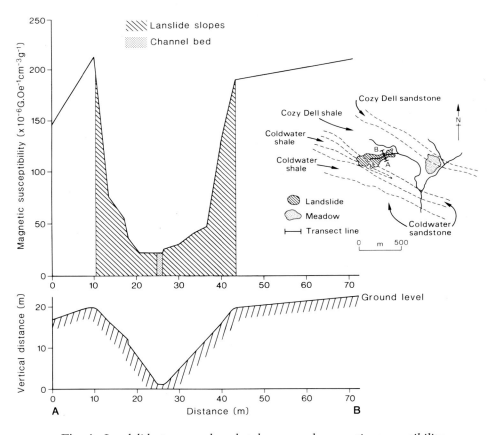

Fig. 4: *Landslide topography, sketch map and magnetic susceptibility.*

slide slopes. In some places the landslide failed to bedrock and the steepness of the slopes and development of a channel along the slide axis have prevented the hollow from becoming infilled with sediment, rather it is acting as an erosional focus. This has caused the low magnetic enhancement values and lack of soil and vegetation development even over a period of at least 56 years. These results show that the spatial variation of magnetic enhancement is partly related to the erosional history of slopes.

3.3 Sediment Storage Sites

Magnetic susceptibility samples were also taken from the sites of sediment storage within the watershed (fig.2). Generally these displayed relatively high susceptibility throughout their depth but little or no surface enhancement. This is because they are formed of soils which have presumably already been enhanced prior to erosion and transport, and continuous deposition has prevented suitable surface conditions for magnetic surface enhancement. An exception to this is the oldest site, the alluvial fan in the Coldwater reach which does display enhancement although it is weak in comparison with the watershed soils. In contrast a bar or terrace fragment within the channel which has small trees growing on it and which is under 100 years old displays no enhancement at all. The sediment from the tunnel which represents suspended sediment deposited in the tunnel during floods has a susceptibility similar to the surface horizons of the Juncal, Matilija and Coldwater soils. The predominance of quartz sand suggests that the principal sources were the Matilija and Coldwater soils. The susceptibility is unlikely to have developed in the tunnel as water runs in it throughout the year producing anaerobic conditions.

3.4 The Role of Fire

Fire is one of the mechanisms responsible for the magnetic enhancement of topsoil (TITE & MULLINS 1971). Fossil evidence from the Santa Barbara Channel shows that fires were frequent during the last 500 years but that in the last 100 years they have increased in frequency and have been less intense and probably less extensive (BYRNNE 1978). The efficiency of fire in inducing magnetic enhancement is dependent upon temperature, soil iron content and organic matter which produces reducing conditions. Soil temperatures in chaparral fires vary greatly due to fuel loading, weather conditions and soil moisture. Of particular relevance to soil properties are the maximum temperatures reached and the duration of heating. Most fires exist at any point for under 30 minutes and temperatures may reach 700°C at the surface, 100-70° at 1 cm, 70°C at 2.5 cm and not alter significantly at 5 cm depth (DE BANO et al. 1979). Preliminary data from experiments using a continuous heat source suggest that the heat pulse travels into chaparral soils at a rate of 5–6 cm hr^{-1} (WELLS, pers. comm.). To assess the effect of heating on the magnetic susceptibility, samples from just below the soil surface (2–4 cm) were burnt under controlled laboratory conditions. The results displayed in fig.5 generally confirm experiments by TITE & MULLINS (1971) and show that enhancement occurs at temperatures above 200°C but does not increase significantly above 500°C. Samples repeatedly burnt at 550°C showed that after the first burn

Soil Development in a Chaparral Watershed

Fig. 5: *Graph showing the results of laboratory ignition of sub-surface samples on magnetic susceptibility.*

Four samples using fresh soil each time show the effect of increasing ignition temperatures on magnetic susceptibility and two samples (broken line) show the effect of cumulative ignition. Lastly one experiment (dotted line) shows the effect of the addition of organic matter on magnetic susceptibility for a sample ignited at 500°C for one hour.

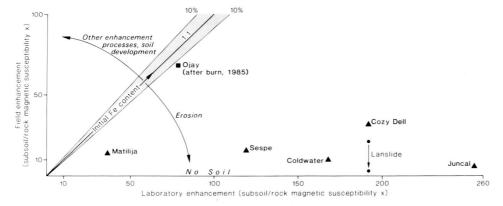

Fig. 6: *Relationship between induced magnetic enhancement by laboratory ignition and measured topsoil enhancement.*

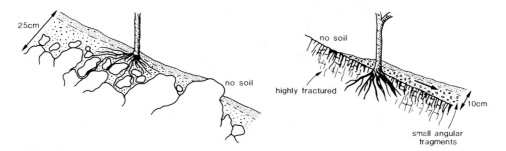

Fig. 7: *Diagrammatic representation of shale (Juncal) and sandstone soil cover in parts of Rattlesnake Canyon.*

little or no further enhancement occurs with subsequent burns. In these experiments the burning atmosphere was not controlled (ie burnt in air) however 5 samples were also burnt in nitrogen then cooled in air as recommended by TITE & MULLINS (1971). Although significantly lower (using a Wilcoxon test) than the values for ignition in air the relative ranking of the samples is the same and three only differ by 15% or less. It is suggested that the reason for high values for ignition in air is that organic matter in the sample and the large sample size required by the magnetic susceptibility equipment (10 g) caused a reducing atmosphere during ignition with oxidation during cooling. Indeed the addition of organic matter (flour) to a non-organic sample produced a regular increase in magnetic susceptibility (fig.5). From tab.2 it can be seen that those samples which showed the highest enhancement after ignition in air also conained the highest quantities of total iron determined by dithionite and pyrophosphate extraction (SOIL SURVEY 1974). Samples from just below the enhanced surface soil were burnt at 550°C for one hour in air to give maximum potential fire induced enhancement and in all cases this was greater than the enhancement measured in the field. When plotted against each other (fig.6) the relative positions of the samples is a measure of potential loss of enhancement due to erosion assuming all were enhanced to maximum values in the 1964 fire. Two samples support this contention, the difference between the very actively eroding middle of the landslide and the edge, and a sample taken just after a wildfire in June of 1985 at Ojay which is within 12% of its laboratory enhanced value. This graph suggests that the Juncal profiles have undergone most rapid erosion and the Matilija soils the least. This is in general agreement with DIBLEE's (1966) assessment of the resistance to erosion of the rock types and is further supported by a much higher drainage density on the Juncal (21 km km^{-2}) than on the Matilija (13 km km^{-2}, KELLER, unpub. data).

4 Conclusions

The low magnetic enhancement and lack of soil depth, low organic matter and lack of soil structure all suggest the Juncal formation is undergoing relatively rapid erosion and transport of the highly fractured but impermeable shale. The few bushes of *Ceonothus* are often rooted in the fractured bedrock and soil moves as a dry-ravel sheet around them. In contrast the higher magnetic susceptibility, greater depth and more highly developed soils on the sandstones and Cozy Dell shale represents a more stable pedogenic situation with lower erosion rates. Fig.7 is a diagrammatic representation of these two situations. The high erodibility of the Juncal is also confirmed by its high drainage density and generally gentler slopes than the rest of the basin.

In contrast sediment storage sites and cumulative soils are characterised by little or no surface enhancement but higher subsurface magnetic susceptibility than contributing soils.

Landsliding is one of the causes of variability in magnetic susceptibility on slopes as it strips off the enhanced horizons exposing subsoil which then undergoes pedogenesis and magnetic susceptibility enhancement unless the slopes become oversteepened preventing soil from forming due to the continuous movement of weathered rock and dry-ravel.

Magnetic susceptibility is useful in a Mediterranean context on non-ferrimagnetic bedrocks for assessing relative soil stability which is inversely related to soil erosion. The role of fire is not only critical in the maintenance of chaparral vegetation but it has important effects on soil properties, erosion and soil development. The common occurrence of surface magnetic susceptibility enhancement of non-Mediterranean soils suggests it may be used in other climatic environments to indicate spatial variations in soil stability, erosion and slope history.

Acknowledgement

The author thanks the Research Board of The University of Leicester for financial assistance and the Geology Department of the University of California, Santa Barbara, Dr. E.A. Keller and other members of the Chaparral Ecosystems Research Group for help and encouragement.

References

BYRNNE, R. (1978): Fossil record discloses wildfire history. California Agriculture **32**, 13–14.

DEARING, J.A., MAHER, B.A. & OLDFIELD, F. (1985): geomorphic linkages between soils and sediments: the role of magnetic measurements. In: Geomorphology and Soils. K.S. Richards, R.R. Arnett & S. Ellis (Eds.) G. Allen and Unwin, London, 245–266.

DE BANO, L.F., RICE, R.M. & CONRAD, C.E. (1979): Soil heating in chaparral fires: Effects on soil properties, plant nutrients, erosion and runoff. USDA Forest Service, Pacific Southwest Frest and Range Experiment Station, Research Paper PSW-**145**.

DIBLEE, T.W. Jr. (1966): Geology of the Central Santa Ynez Mountains, Santa Barbara County, California. Division of Mines and Geology, Bulletin **186**.

JOHNSON, D.L. (1985): Soil and thickness processes. In: P.D. Jungerius (Ed.), Soils and geomorphology. CATENA SUPPLEMENT **6**, 29–40.

KELLER, E.A. et al. (1985): Rattlesnake Canyon. Chaparral Watershed Group. University of California, Santa Barbara. Unpublished report.

LE BOGNE, E. (1955): Abnormal magnetic susceptibility of the topsoil. Annls. Geophys. **11**, 399–419.

MULLINS, C.E. (1977): Magnetic susceptibility of the soil and its significance in soil science — a review. J. Soil Sci. **28**, 223–246.

MUTCH, R.W. (1970): Wildland fires and ecosystems — A hypothesis. Ecology **51**, 1046–1051.

RICE, R.M., ZIEMER, R.R. & HANKIN, S.C. (1982): Slope stability effects on fuel management strategies — inferences from Monte Carlo simulations. In: Procs. Symp. on Dynamics and Management of Mediterranean Type Ecosystems; June 1981, San Diego, CA. Gen. Tech. PSW-**58**. Washington DC: Forest Service US Dept. Agric., 365–371.

SOIL SURVEY (1974): Laboratory Methods. Technical Monograph **6**, Harpenden, England.

TAYLOR, B.D. (1983): Sediment yields in coastal southern California. J. Hydrol. Eng. ASCE, **109**, 71–85.

TAYLOR, R.M. & SCHWERTMANN, U. (1974): Maghemite in soils and its origin. II Maghemite synthesis at ambient temperature and pH 7. Clay Mins. **10**, 299–310.

TITE, M.S. & LININGTON, R.E. (1975): Effect of climate on the magnetic susceptibility of soils. Nature, London, **256**, 565–566.

TITE, M.S. & MULLINS, C. (1971): Enhancement of the magnetic susceptibility of soils on archaeological sites. Archaeometry **13**, 209–219.

Address of author:
Dr. A.G. Brown
Geography Department, The University
Leicester LE17RH
England

SEASONALITY OF SUBSURFACE FLOW AND NITRATE LEACHING

T.P. **Burt**, Oxford

Summary

The generation of subsurface runoff (throughflow) in a small drainage basin in a humid temperate environment is shown to be strongly seasonal. Lateral downslopw flow is largely confined to the winter period (December to February inclusive); recharge of the soil moisture deficit during autumn is necessary before this can be achieved. The pattern of nitrate leaching from the drainage basin is also strongly seasonal, being controlled by the occurrence of subsurface runoff, with three-quarters of the annual nitrate loss in the winter.

Resumen

La formación de una circulación lateral subsuperfical (throughflow) de agua en las vertientes de en una pequeña cuenca de drenaje, situada en un medio templado húmedo, tiene un carácter marcadamente estacional. Este flujo a lo largo de las laderas se produce en su mayor parte durante el período invernal (diciembre a febrero inclusives); sin embargo, antes de que esto suceda, es necesario que durante el otoño se haya producido una recarga del déficit de humedad del suelo. El lavado de nitratos de la cuenca es también marcadamente estacional, y debe de estar controlado por esta escorrentia subsuperficial puesto que las tres cuartas partes de la pérdida anual de nitrato tiene lugar en invierno.

1 Introduction

Strong seasonal contrasts in the operation of geomorphological processes are not normally associated with humid temperate environments. However, with respect to water movement within the soil, rates of subsurface flow are likely to differ markedly between summer and winter. For land with impermeable subsoil or bedrock, soil water recharge in autumn will usually be sufficient to reestablish lateral flow within the soil (throughflow) towards the local stream or tile drain, except under drought conditions when the winter is too dry. This paper demonstrates the strong seasonality of subsurface runoff for a small drainage basin in a humid temperate environment, and examines the implications of this pattern of flow in relation to nitrate leaching.

2 The Study Site

The Slapton Wood catchment in south Devon, England, has an area of 94 ha. and is a second order tributary of the

River Gara which drains into Slapton Ley, the largest natural body of freshwater in south-west England and an important nature reserve. Hydrometric measurements have been made since 1969 when concern over increasing eutrophication led to the need to establish runoff and nutrient inputs from all of the streams draining into the lake. The early results of this research were summarised by TROAKE & WALLING (1973) and by TROAKE et al. (1976). Over the last decade the Slapton Wood stream has received particular attention, not just in relation to water and solute budgets (BURT et al. 1988): hillslope runoff processes have been studied (BURT & BUTCHER 1985a, b) and related to the source areas for runoff and solutes (BURT & ARKELL 1986).

The topography of the Slapton Wood catchment can be divided between large plateaus of low gradient (where most of the arable farming takes place) above steep valleyside slopes which reach a maximum angle of 25 degrees. The plateaus drain to the main stream via three large hillslope hollows, one of which has been the site for extensive soil moisture investigations (BUTCHER 1985) using tensiometers and piezometers. Stream discharge leaving the basin is measured using a 120-degree thin-plate V-notch weir and Ott stage recorder; automatic water samplers are used at this site to collect samples for analysis of sediment and solute concentrations. Elsewhere within the basin, the use of weirs or flumes, together with water samplers, has allowed the spatial pattern of runoff and sediment production to be identified (BURT & ARKELL 1986). The soils are permeable, acid, nutrient-poor, silty clay loams of the Denbigh series (TRUDGILL 1983) over an impermeable bedrock of Devonian shales and slates. A mean infiltration capacity of 9 mm hr-1 has been noted, but the soils are easily compacted by animals or farm vehicles and infiltration capacity may fall to 2 mm hr-1. COLES & TRUDGILL (1985) have identified a pedal infiltration capacity of about 2.5 mm hr-1. with preferential flow occurring through macropores at higher rainfall intensities. Overland flow is mainly confined to variable source areas at the base of hillslope hollows, but widespread infiltration-excess overland flow is not unknown when soil and rainfall conditions are suitable. As noted below, most of the basin runoff occurs as subsurface flow. Mean annual rainfall is 1035 mm with a mean annual temperature of 10.5°C.

Land use in the Slapton Wood catchment is a mixture of permanent pasture and arable (barley, fodder crops and vegetables) and only 13% of the basin is in fact wooded. Fertiliser use averaged 46 kg ha-1 nitrate-nitrogen (NO_3-N) in 1983; such usage is relatively modest compared to figures reported elsewhere in the United Kingdom (ROYAL SOCIETY 1983). Even so, the annual loss of N from the catchment ranges between 50% and 105% of the annual fertiliser application (TROAKE et al. 1976, BURT et al. 1988). This loss is highly dependent on the production of subsurface runoff and is consequently strongly seasonal in its occurrence.

3 Temporal Variations in Runoff and Nitrate Yield

Tab.1 shows monthly totals of rainfall, stream discharge and nitrate load, together with the mean monthly nitrate concentration, for the Slapton Wood

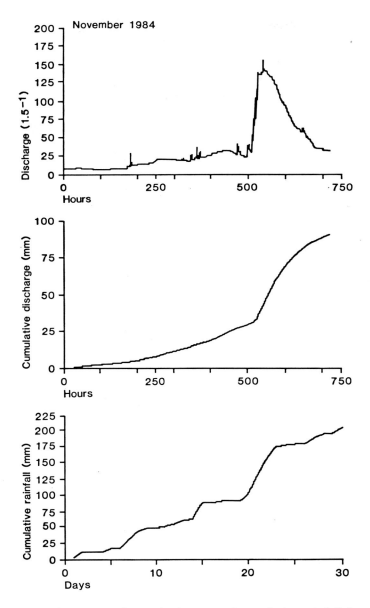

Fig. 1: *Stream discharge, cumulative discharge, and cumulative rainfall for the Slapton Wood stream, November 1984.*

	Month	Rainfall (mm)	Discharge (mm)	Mean NO_3-N concentration (mg l-1)	Nitrate load (kg NO_3-N)
1983	October	75	12.0	6.37	71.86
	November	69	12.6	7.00	82.25
	December	117	83.0	7.75	604.72
1984	January	243	184.0	7.79	1346.74
	February	70	104.0	9.15	894.58
	March	65	29.0	8.29	225.71
	April	8	24.7	7.05	163.44
	May	62	16.2	6.62	100.63
	June	9	8.4	6.19	48.88
	July	34	5.5	5.86	130.28
	August	54	5.3	6.01	29.83
	September	89	4.1	7.17	27.95
	October	123	10.4	6.65	65.21
	November	203	95.7	9.72	874.12
	December	106	100.8	10.26	971.38
1985	January	114	114.4	9.81	1055.00
	February	54	99.8	9.99	937.86
	March	92	28.7	9.18	247.26
	April	65	70.7	8.62	572.11

Tab. 1: *Monthly totals of discharge and nitrate load for the Slapton Wood catchment for the period Ostober 1983 to April 1985.*

stream for the period October 1983 to April 1985. These data show that the major loss of nitrate occurs in winter when both discharge and nitrate concentration are at their highest. In the 1984 water year (beginning 1 October 1983), 79% of the total nitrate load was exported from the basin in the three months, December to February inclusive. The annual loss of nitrate is equivalent to about 80% of the annual fertiliser input of nitrate to the basin.

Only 1% of the annual runoff in the Slapton Wood stream is quickflow (TROAKE & WALLING 1973). Not all of this is surface runoff: some is rapid subsurface flow from saturated areas close to the stream, or via macropores. The remaining 99% is entirely generating from within the soil as throughflow, since the bedrock is impermeable.

Much of the throughflow occurs in winter in the form of delayed hydrographs which peak up to several days after the rainfall, with high flows lasting for as long as two weaks. BURT & BUTCHER (1986) examined 44 such hydrogaphs for the period 1969–1982. They found that during the winter half of the year, these hydrographs occur for 54% of the time and contribute 46% of the total runoff; by contrast quickflow occupies only 2% of the time, producing less than 4% of the runoff. In addition, as noted below, delayed hydrographs are times of major nitrate loss since both discharge and nitrate concentration are high. The generation of the delayed hydrograph is associated with the development of a large "wedge" of saturated soil which builds up above the soil-bedrock interface. The saturated wedge extends upslope from the stream and, during periods of high subsurface flow, it is continuous right on to the plateau areas above the steeper valleyside slopes. Drainage from the plateau is thought to be particularly important in generating the delayed hydrograph peak; much of the drainage is via the large hillslope hollows described above (BURT & BUTCHER 1985a, b). These large delayed hydrographs occur only in the winter months when the soil water deficit has been recharged. This is shown clearly on fig.1: this shows rainfall and runoff for November 1984 which was a particularly wet month with 203 mm rainfall. The plot of cumulative discharge shows a less marked gradient than that of cumulative rainfall. Thus, the rainfall early in the month went to recharge soil moisture rather than to produce runoff. Once soil moisture levels are sufficiently augmented, even a small amount of rainfall can trigger a large subsurface response. This occurs after 18 November: cumulative discharge rapidly increases from that time. The discharge record shows that the first major delayed hydrograph of the winter peaks on November 23 (though two small delayed hydrographs are detectable earlier in the month).

Since the study period covers two winters, it is of some interest to compare nitrate losses. The 1983/84 winter was relatively dry; not only was the next winter wetter, but it followed a summer drought. Nitrate concentrations and loads were both very much higher as a result, a feature which was also noted in many catchments at the end of the 1975/76 drought (see below). For the winter half-year October to March, rainfall in 1983/84 was 639 mm compared to 692 mm in 1984/85; runoff was similar (425 mm compared to 450 mm); but nitrate losses in the second winter were almost one metric ton greater (3226 kg compared to 4151 kg). Two factors help to explain this difference: much organic matter would have been mineralised and made available for leaching during the hot summer of 1984; in addition, lack of effective leaching the previous winter will also have led to the accumulation of nitrate in the soil. It is also very relevant to note that there were only 4 delayed hydrographs during the drier 1983/84 winter, whilst there were 9 such events in the 1984/85 winter. The delayed hydrograph is a particularly important time for nitrate leaching: saturation at the base of the soil profile is produced by soil water moving down from the root zone, leaching nitrate as it does so; in addition, growth of the zone of saturation towards the soil surface may cause more effective removal of nitrate from surface horizons than can occur with unsaturated flow. Both these processes combine to produce high nitrate concentrations in drainage waters at a time of high flow: the result is a major loss of nitrate from the soil (BURT et al. 1983). The significance of delayed hydrographs is further confirmed by reference to the 1983/84 winter: the 4 hydrographs noted above were responsible for 70% of the nitrate loss in the period December to February.

Thus, the loss of nitrate is strongly seasonal and closely linked to the generation of delayed throughflow hydrographs. Winter losses of nitrate are crucial, not just because of the total amounts lost, but also because nitrate concentrations are highest at this time. The World Health Organisation recommended limit for nitrate in drinking water is 11.3 mgl^{-1} NO$_3$-N. Nitrate-nitrogen concentrations in the Slapton

Wood stream have been steadily rising over the last 15 years at about 0.2 mgl^{-1} NO$_3$-N per annum (BURT et al. 1988). In wet winters, particularly those following drought summers, NO$_3$-N levels may now exceed 11.3 mgl^{-1}. In the 1983/84 winter, NO$_3$-N concentrations were above 10 mgl^{-1} for much of February, but never reached 11.3 mgl^{-1}. By contrast, in the 1984/85 winter, NO$_3$-N concentrations were consistently higher and exceeded 11.3 mgl^{-1} for 8 days at the basin outlet (Upstream, where the stream leaves the agricultural area and enters the wood, the WHO limit was exceeded for 57 days). The winter of 1976/77 was also wet and followed a major drought: significant losses of nitrate, at very high concentrations, were observed then too, both in the Slapton Wood catchment (BURT et al. 1988), and elsewhere (FOSTER & WALLING 1978).

4 Conclusions

Even in a humid temperate region, lack of summer rainfall in combination with modest evaporation produces a significant soil moisture deficit. Subsurface runoff does not occur in autumn until the soil moisture deficit has been removed. The result is that nitrate leaching is strongly seasonal, being confined to the winter period when up to 80% of the annual nitrate loss from the catchment may take place. Wet winters, and those following major droughts, are likely to be characterised by major losses of nitrate from the catchment, with higher nitrate concentrations than is normal. The results suggests that the practice of autumn sowing of crops should be encouraged: any means of retaining nitrate in the soil during the winter (ie by plant uptake) will reduce the potential for leaching. The discussion also implies that fields adjacent to the steep valleyside slopes, particularly those draining to hillslope hollows, are more susceptible to leaching than other fields on the plateau nearer the watershed. It may be that such areas require greater attention, and the fertilisers should be most carefully used at such sites.

Although much research has studied soil moisture patterns on slopes and leaching processes within the soil profile, much remains to be done if the precise hydrological pathways involved in the **lateral** removal of nitrate from hillslopes is to be fully understood. Further work is being carried out at Slapton to trace the movement of soil water downslope. This should define more closely the mechanisms involved in the generation of delayed hydrographs, and should identify links between slope drainage and nitrate loss. This knowledge may then be helpful in devising land use strategies to minimise winter losses of nitrate from catchments.

References

BURT, T.P. & BUTCHER, D.P. (1985a): Topographic controls of soil moisture distributions. Journal of Soil Science, **36**, 469–486.

BURT, T.P. & BUTCHER, D.P. (1985b): On the generation of delayed peaks in stream discharge. Journal of Hydrology, **78**, 361–378.

BURT, T.P. & ARKELL, B.P. (1986): Variable ource areas of stream discharge and their relationship to point and non-point sources of nitrate pollution. International Association of Hydrological Sciences Publication **157**, 155–164.

BURT, T.P. & BUTCHER, D.P. (1986): Development of topographic indices for use in semi-distributed hillslope runoff models. Zeitschrift für Geomorphologie, Supplementband **58**, 1–19.

BURT, T.P., BUTCHER, D.P., COLES, N. & THOMAS, A.D. (1983): The natural history of the Slapton Ley nature reserve. XV. Hydrological processes in the Salpton Wood catchment. Field Studies, **5**, 731–752.

BURT, T.P., ARKELL, B.P., TRUDGILL, S.T. & WALLING, D.E.(1988): Stream nitrate levels in a small catchment in south west England over a period of 15 years (1970–1985). Hydrological Processes, in press.

COLES, N. & TRUDGILL, S.T. (1985): The movement of nitrate fertiliser from the soil surface to drainage waters by preferential flow in weakly structured soils, Slapton. S. Devon. Agriculture, Ecosystems and Environment, **13**, 241–259.

FOSTER, I.D.L. & WALLING, D.E. (1978): The effects of the 1976 drought and autumn rainfall on stream solute levels. Earth Surface Processes, **3**, 393–406.

ROYAL SOCIETY (1983): The nitrogen cycle of the United Kingdom. Royal Society, London.

TROAKE, R.P. & WALLING, D.E. (1973): The natural history of the Slapton Ley nature reserve. VII. The hydrology of the Slapton Wood stream. Field Studies, **3**, 719–740.

TROAKE, R.P., TROAKE, L.E. & WALLING, D.E. (1976): Nitrate loads in south Devon streams. MAFF Technical Bulletin **32**, 340–351.

TRUDGILL, S.T. (1983): The natural history of the Slapton Ley nature reserve. XVI. The soils of Slapton Wood. Field Studies, **5**, 833–840.

Address of author:
T.P. Burt
School of Geography, University of Oxford
Mansfield Road
Oxford OX1 3TB
UK

MEASUREMENTS OF CAVERNOUS WEATHERING AT MACHTESH HAGADOL (NEGEV, ISRAEL) A SEMIQUANTITATIVE STUDY

K. Rögner, Trier

Summary

Observations and measurements carried out in tafoni in Machtesh HaGadol (Negev, Israel) demonstrate active flaking and cavernous weathering. Weathering is dependent upon the prevailing arid climate, however, the most important control is the occurence of moisture. The evaporation of moisture with highsaturated or supersaturated dissolved load of salts results in a progressive build up of interstitial salt horizons parallel to the cavern of the tafoni behind the developing flakes.

Rates of cavernous weathering measured by different methods reach 0.04 mm to 0.2/0.3 mm/year.

Zusammenfassung

Beobachtungen und Messungen an verschiedenen Tafoni im Machtesh Ha-Gadol (Negev, Israel) zeigen, daß das Abschuppen im Inneren der Tafoni und somit der Prozeß der Tafonierung (kavernöse Verwitterung) aktive morphodynamische Vorgänge sind, die in Einklang mit dem heute herrschenden ariden Klima stehen. Wichtigste Voraussetzung ist das Vorhandensein von Feuchtigkeit, denn die Verdunstung der mit gelösten Salzen beladenen Feuchte führt zu einer fortschreitenden Anreicherung der Salze in Form von Interstitialfüllungen parallel zum Tafonehohlraum hinter den sich bildenden Schuppen. Die mit Hilfe verschiedener Methoden gewonnenen Werte über die Intensität der Abschuppung und damit der kavernösen Verwitterung erreichen Beträge zwischen 0.04 und 0.2/0.3 mm/Jahr.

Resumen

Las observaciones y mediciones llevadas a cabo en los tafonis de Machtesh Ha-Gadol (Negev, Israel) demuestran una escamación y una meteorización en cavernas muy activas. En un clima árido el factor de control más importante en la meteorización es la ocurrencia de humedad. La evaporación, cuando existe un contenido saturado o supersaturado en sales disueltas, tiene como resultado la progresiva acumulación intersticial de horizontes salinos paralelos a las cavernas de los tafonis y detrás de las escamas en desarrollo. Las tasas esta meteorización en cavernas medida con diferentes métodos varia entre los 0.04 mm a los 0.2/0.3 mm/año.

ISSN 0722-0723
ISBN 3-923381-12-3
ⓒ1988 by CATENA VERLAG,
D-3302 Cremlingen-Destedt, W. Germany
3-923381-12-3/88/5011851/US$ 2.00 + 0.25

1 Introduction

Cavernous weathering was originally described during the last century from the Mediterranean area (1878 for the first time by REUSCH; quoted in KVELBERG & POPOFF 1937, 131–137 where some illustrations of REUSCH are also reprinted). A. PENCK (1894; quoted in KVELBERG & POPOFF 1937, 139) proposed naming these forms of cavernous weathering 'tafoni' (sing. form: tafone). This term is said to be based on the Corsican word 'tafonare' which means to be perforated, and the inhabitants of Corsica call "every boulder of stone and also the outcropping rock which is perforated with cavities as 'pietra tafonata' " (WILHELMY 1981, 155).

Later on the term tafoni was extended to cavernous forms outside of the Mediterranean climate and to lithologies other than crystalline rocks (LOUIS & FISCHER 1979, 133, OLLIER 1979, 191, WILHELMY 1981, 143).

However, even today one can still find the view expressed that tafoni are typical weathering forms of the Mediterranean subtropic zone (for a definition of that climate see ROTHER 1984) or in similar climates. If tafoni occur in other climatic regions, then they should, according to this viewpoint logically be relict or "morphodynamically inactive" features. If, for example, tafoni are found in dry arid regions, they should indicate wetter paleoclimatic conditions. In that case the tafoni would be forms which are no longer actively developing in arid climates, they would be morphodynamically inactive ('dead') forms in arid environments (see the comprehensive study at the Ayers Rock in Central Australia by BREMER 1965). The tafoni in the Negev Desert of southern Israel (fig.1) and especially those of Machtesh Ha-Gadol are also thought to be forms of wetter paleoclimates (BREMER 1975).

The latter statement in particular, contrasted with our initial impressions (1974) obtained at Machtesh HaGadol.

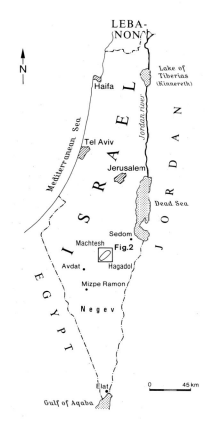

Fig. 1: *Location map.*

In order to investigate in more detail this contradiction to our own observations — which seemed to be identical with the results of KVELBERG & POPOFF (1937) and KLAER (1956) from the mediterranean region (Corsica) — we have studied between 1976 and 1982 numerous tafoni in Machtesh Ha-

			Reference
Temperature (Machtesh HaGadol)			
annual mean		21°C	1
hottest month		26–28°	1
coolest month		10°C	1
absolute maximum (Advat)		46.4°	2
absolute minimum (Advat)		−4.5°	2
Precipitation (Advat)			
annual mean		90 mm	1
dry year		25.6 mm	3
moist year		152.7 mm	3
Months without rain at Sedom and Mizpe Ramon		June, July, August, September	4
Evaporation, (Advat, class "A" pan)			
annual mean		2615 mm	5
July		314.4 mm	5
January		96.5 mm	5
Dewfall (Advat)			
annual mean;	surface	32.53 mm	6
	1 m	31.07 mm	6
dry year;	surface	24 mm	6
	1 m	22 mm	6
moist year;	surface	42 mm	6
	1 m	42 mm	6
average;	moist month	4.7 mm	7
	dry month	1.0 mm	7
Relative humidity (Machtesh HaGadol)			
annual mean		50–55%	1
August 1979		18–100%	8
March 1982		20–95%	8

References:
1) ATLAS OF ISRAEL (1970)
2) EVENARI (1982, 22)
3) EVENARI (1982, 80/81)
4) Own computations based on vols. 1–33 (1950–1982) of Statistical Abstract of Israel
5) EVENARI (1982, 22)
6) EVENARI (1982, 7)
7) EVENARI et al. (1971, 36)
7) Own measurements

Tab. 1: *The climatic environment in the Machtesh HaGadol area (Negev, Israel).*

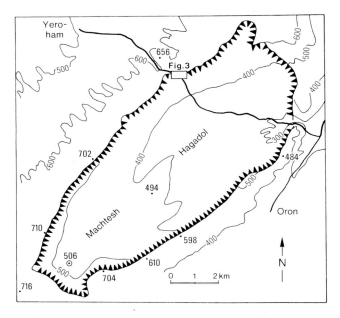

Fig. 2: *Machtesh HaGadol in the Northern Negev of Israel with the measurement area (fig.3) at the foot of the surrounding cliff. Contour lines are given in intervals of 100 m; points with numbers indicate the elevation.*

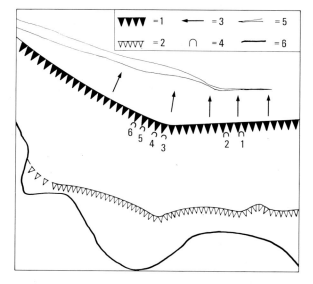

Fig. 3: *The measurement area of Machtesh HaGadol showing the location of the studied tafoni.*

1 = cliff of Machtesh; 2 = step with heigh of 5 m; 3 = direction of surface runoff above the cliff; 4 = location of the tafoni; 5 = small wadi; 6 = road.

Gadol. Six of these were analysed in detail between 1977 and 1982 in order to obtain information on rates of flaking and scaling. The results of the analysis are presented in this paper and they complement an earlier study of the microclimatic environment and its influence on cavernous weathering (RÖGNER 1987).

1.1 Location and Climate

The Machtesh HaGadol lies in the center of the Northern Negev (fig.1) in southern Israel. The climate is arid with strong seasonal contrasts (tab.1) caused by the influence of rain-bearing cyclones during winter and by the presence of the subtropical anticyclone belt during summer.

2 Methods

From 1977 to 1982 the weathering rates of six different tafoni in Machtesh HaGadol (fig.2, 3) were measured using a range of methods. Initially, plastic sheeting was placed on the floors of the tafoni to collect flakes and scales. This method was, however, not generally practicable, because the plastic was destroyed and blown by strong winds.

Later, coloured stripes were spray-painted across the roof and the backwall of one tafoni (1977) and then in six further tafoni (1979–1982).

Finally in 1979 sampling tins (cake tins with high rims) were set out to collect the flaked and scaled material in the center of three tafoni.

To avoid confusion flaking is defined as the weathering of rocks into flakes between 1 and 10 mm thick, which form parallel to the rock surface, and scaling as the formation of scales between 1 and 10 cm thick (SMITH 1978, 26).

Most of the collected rock pieces were flakes and often the thickness was <1 mm. Field observations and the material from the sampling tin suggest that the often mentioned granular disintegration is insignificant in comparison with the dominant process of flaking. The markings and also the sampling tins were examined in September 1980, March 1981, March 1982 and July 1982, using the following procedures.

One: Fallen flakes and scales with coloured markings were collected inside and outside of the tafoni, separated on the basis of colours (spray-painted colours; 1977: orange, 1979: red, 1980: blue, 1981: black) and weighed. By using the specific weight of the fallen material and relating it to the interior surface area of the tafoni (without floor) a value for the average annual rate of cavernous weathering per area unit was determined. Since material also flakes (and scales) off unmarked areas, method 1 can be used only for calculating minimum values.

Two: The length and the width of the coloured stripes were measured and the square dimensions (length × width) calculated. Then the hand-collected material (separated on the basis of colour) and the specific weight of the outcropping rock are used to calculate the average flaking rates for the different tafoni. On the condition that all of the flaked material with coloured marking can be sampled, method 2 results in overestimations because the flaked material is mostly only partly marked.

Three: As in method 2, the length and width of the coloured stripes was measured after spraying. On each visit the painted and flaked areas of the stripes were measured and compared. Using measurements of the flake thickness the range and average rate of flaking could be calculated. Again, this method includes only the marked areas of the tafoni. An analysis which covers the whole area of the tafoni cavern is nearly impossible.

Four: On each visit fallen flakes were collected from sampling tins with high rims installed in the center of the tafoni for collecting the material which had scaled off directly above the tins. The large size of the flakes and scales normally excluded any addition by eolian processes.

Unfortunately one of the sampling tins disappeared between 1979 and 1981 and another between 1981 and 1982. For this reason long-term sampling was only possible at tafone number two. However, holes drilled in the roof during microclimatical studies (RÖGNER 1987), perhaps explain this high value.

An experiment involving the comparison of photographs taken from the same point at different times was generally not successful, as scaling could be observed on coloured (natural or artificial) areas but not on surfaces where the stone colour is nearly white — as is largely the case at tafone 'four' where the photography experiment was carried out.

Unfortunately this tafone had to be chosen because the roofs of the other tafoni were unsuitable for photographing.

3 Results

Flaking and scaling, which were initially observed and later measured, are the principle causes of cavernous weathering and therefore also for the genesis of the tafoni in Machtesh HaGadol.

A comparison of values for flaking/scaling (tab.2) based on the different methods of assessment shows that the values obtained by method 1 are consistently lower than those by method 2, by a factor of approximately ten. This result was predicted in the description of the methods, because method 1 yields only minimum values. The results obtained by methods 2 and 3 do not differ greatly, whereas method 4 yielded the highest values at tafone 'two' but not at tafone 'six'. If the 'minimum values' of method 1 are neglected, scaling/flaking rates were (and with these also rates of tafoni formation) between 0.04 mm/year (tafone 'four', August 1979 – March 1981; method 2) and 0.7 mm/year (tafone 'two', March 1981 – March 1982, method 4) using methods 2, 3 and 4.

Field observations suggest that granular disintegration is insignificant in comparison with flaking/scaling.

As the three highest values were recorded without exception at tafone 'two' where holes for thermocouples had been drilled, the measured tafonisation values can be assumed to be somewhat too high. A maximum rate of 0.2–0.3 mm/year for cavernous weathering is therefore more realistic than the higher rate. The measured values do not only differ for reasons related to the methods. They fluctuate at the same tafoni depending on the different periods of measurement and, naturally, between the different tafoni.

Because the tafoni are situated in different rocks (tafone 'four' in limestone strata, the others in dolomitic strata) the first impression is that the different rates of flaking/scaling are related to rock type. But this supposition cannot be confirmed by the results presented in this paper.

The different chemical composition seems to be less responsible for the varying levels of intensity than the ability of

Measurements of Cavernous Weathering

tafone	time	\multicolumn{4}{c}{rates of cavernous weathering during 1 year according to}			
		method 1	method 2	method 3	method 4
1	Sept. 1980–March 1982	0.008 mm	0.066 mm		
2	Aug. 1979–March 1981				0.34 mm/a
	March 1981–March 1982	0.05 mm			0.7 mm/a
	Aug. 1979–July 1982				0.6 mm/a
3	Sept. 1980–March 1982	0.01 mm	0.08 mm		
	March 1981–March 1982	0.03 mm	0.03 mm*		
	8.3.1982–28.7.1982	0.06 mm	0.21 mm		
4	Sept. 1977–Aug. 1979			0.05 – 0.2 mm	
	Aug. 1979–March 1981	0.004 mm	0.04 mm	0.075 – 0.3 mm	
	March 1981–March 1982	0.02 mm	0.1 mm	0.075 – 0.3 mm	
	8.3.1982–28.7.1982	0.01 mm	0.08 mm		
5	Aug. 1979–March 1981	0.008 mm	0.27 mm		
	March 1981–March 1982	0.003 mm	0.06 mm		
	Aug. 1979–March 1982		0.19 mm		
6	Aug. 1979–March 1981				0.1 mm

* This calculation is based only on the coloured marking of 1979, the marking of 1981 is missing. The result has a value which is too small.

Tab. 2: *The rates of cavernous weathering calculated for one year.*

tafone	minimum rate in mm/year	maximum rate in mm/year	method
2	August 1979–March 1981 (0.34 mm/a)	March 1981–March 1982 (0.7 mm/a)	4
3	Sept. 1980–March 1981 (0.01 mm/a)	8.3.1982–28.7.1982 (0.06 mm/a)	1
	Sept. 1980–March 1981 (0.08 mm/a)	8.3.1982–28.7.1982 (0.21 mm/a)	2
4	August 1979–March 1981 (0.004 mm/a)	March 1981–March 1982 (0.02 mm/a)	1
	August 1979–March 1981 (0.04 mm/a)	March 1981–March 1982 (0.1 mm/a)	2
	September 1977–Aug. 1979 (0.05 – 0.2 mm/a)	August 1979–March 1981 + March 1981–March 1982 (0.075 – 0.3 mm/a)	3
5	March 1981–March 1982 (0.003 mm/a)	August 1979–March 1981 (0.008 mm/a)	1
	March 1981–March 1982 (0.06 mm/a)	August 1979–March 1981 (0.27 mm/a)	2

Tab. 3: *The cleared maximum and minimum rates of cavernous weathering.*

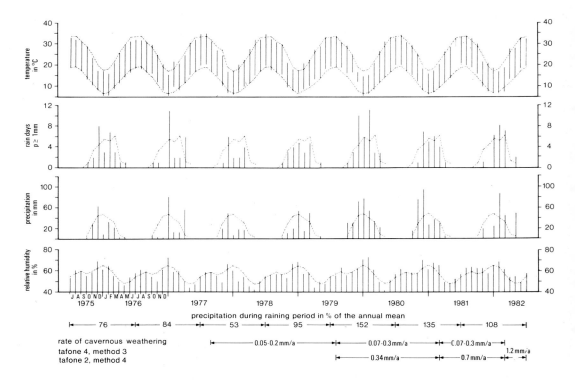

Fig. 4: *The climatic conditions in Ber Sheva during the periods of measurement and the rates of cavernous weathering at Machtesh HaGadol.*

the rock strata to hinder or favour seepage of rock moisture. A rock stratum which permits or favours seepage as a result of sufficient pore space or pore size will show more intensive flaking and scaling than a rock stratum which handicaps the seepage processes (RÖGNER 1985). This is the result of observations in other regions with a different petrographic setting (Elat region, Israel; Sinai). Furthermore the different tafoni are not at the same stage of development. In other words, the tafoni occupy different positions in the morphogenetic sequence of tafonisation (EICHLER & RÖGNER 1978, RÖGNER 1985).

Tafone 'one', for example, is not fully developed because a well developed roof is missing; tafoni 'two' and 'six' show the ideal development of a tafone cavern with a well-devloped roof. Tafoni 'three' and 'four' have passed the ideal stage and parts of their roof have collapsed, whilst at tafone 'five' only a very thin roof is left in which seepage processes cannot operate and flaking is only active on the backwall.

Field observations carried out at many more then the six above mentioned tafoni have indicated that well-developed tafoni produce more flakes and scales than the less-developed or nascent tafoni.

Finally, if the minimum and maximum values, as well as the different time spans of the observations and measurements are taken into account it appears possible that the intensity of flaking/scaling can also to a certain extent be influenced by the mesoclimate. For example most minimum weathering values (tab.3) occured between August 1979 and March 1981, a time span which also includes the period of September 1980 to March 1981 (tafone 'three'). The value for the period September 1977 to August 1979 is not very reliable because the measurements were only carried out at tafone 'four' (method 3).

Nearly all of the maximum weathering values were measured from March 1981 to March 1982 and from March 8, 1982 to July 28, 1982. Only tafone 'five' differs from this pattern and gave minimum values during the latter time span, and vice versa. The difference probably can be explained by the fact that several huge scales broke away from the backwall of the cavern between August 1979 and March 1981, while such processes did not occur between March 1981 to March 1982.

If the different rates of cavernous weathering are evaluated with respect to the monthly rainfall data, the number of rain days, the minimum and maximum temperatures and the average humidity (fig.4) it can be concluded that:

- There are no evident relationships between the intensity of flaking/scaling and the monthly minimum and maximum temperatures, nor the number of days.

- Higher precipitation seems to influence flaking with a delay of one to two years; higher relative humidity with a delay of only one year. Higher precipitation can be assumed to result in higher amounts of rock moisture, higher relative humidity in an increase of moisture from outside the cavern.

- Higher rates of cavernous weathering occur without any delay during the dryer periods since flaking seems to be greatest when limited moisture allows more frequent wetting/drying and associated weathering such as salt weathering.

The evaporation of moisture (from different sources like rock moisture, relative humidity, dewfall, rain) with a high saturated or supersaturated dissolved load of salts results in a progressive build up of interstitial salt horizons. The changes in rock surface temperature and relative humidity combined with the evaporation result in repeated stressing behind the developing flakes.

The results and their evaluation demonstrate that flaking, scaling, and consequently cavernous weathering are active morphodynamic processes in the arid Negev, which are closely related to the present climate. The necessary condition is the occurence of moisture as rock moisture, as relative humidity, as dew- or rainfall. The intensity of flaking, scaling and cavernous weathering attains rates of 0.04 mm/year to 0.7 mm/year, whereby the latter seems to be too high and a rate of 0.2 to 0.3 mm/year could be a realistic maximum.

Acknowledgement

I would like to thank Dr. H. Eichler for his assistance and friendly help during the fieldwork in the Negev.

Dr. Zapf (Paderborn) and Dr. B.J. Smith (Belfast) helped with the preparation of the translation.

The German Research Fund (DFG) has sponsored three of the five field studies.

References

ATLAS OF ISRAEL (1970): Survey of Israel, Ministry of Labour, Jerusalem and Elsevier Publishing Company, Amsterdam.

BREMER, H. (1965): Ayers Rock, ein Beispiel für klimagenetische Geomorphologie. Zeitschrift für Geomorphologie N,F. **9**, 249–284.

BREMER, H. (1975): Intramontane Ebenen, Prozesse der Flächenbildung. Zeitschrift für Geomorphologie N.F. Supplementband **23**, 26–48.

EICHLER, H. & RÖGNER, K. (1978): Holozäne Morphodynamik und Intensitätsstufen des Kleinformenschatzes im vollariden Südosten Israels. Zeitschrift für Geomorphologie N.F. Supplementband **30**, 162–182.

EVENARI, M. (1982): Ökologisch-landwirtschaftliche Forschungen im Negev — Analyse eines Wüsten-Ökosystems. Darmstadt (Technische Hochschule).

EVENARI, M., SHANAN, L. & TADMOR, N. (1971): The Negev, the challenge of a desert. Cambridge Mass. (Harvard Press).

KLAER, W. (1956): Verwitterungsformen im Granit auf Korsika. Petermanns Geographische Mitteilungen Ergänzungshefte **261**. Gotha.

KLAER, W. & WASCHBISCH, R. (1981): Neuere Erkenntnisse über den Prozeß der Tafoniverwitterung. Aachener Geographische Arbeiten **14**, 67–79.

KVELBERG, I. & POPOFF, B. (1937): Die Tafoni-Verwitterungserscheinung. Petrographische Beiträge zur Aufklärung des Tafoniproblems. Latvijas Universitates Ruksti (Acta Universitatis Latviensis) Kimijas Fakultates Serija **IV.6**, 125–368. Riga.

LOUIS, H. & FISCHER, K. (1979): Allgemeine Geomorphologie. Lehrbuch der Allgemeinen Geographie **1**. Berlin — New York.

OLLIER, C. (1979): Weathering. London — New York.

RÖGNER, K. (1985): Geomorphologische Untersuchungen im Negev und Sinai. Ein Beitrag zur rezenten Morphodynamik, zur Quantifizierung arid-morphodynamischer Prozesse und Prozeßkombinationen sowie zur Landschaftsgenese. Manuskript der Habilitationsschrift (Universität Paderborn).

RÖGNER, K. (1987): Temperature measurements of rock surfaces in hot deserts (Negev, Israel). International Geomorphology 1986 (= Proceedings of the First International Conference on Geomorphology), Part II, 1271–1287. Chichester.

ROTHER, K. (1984): Mediterrane Subtropen. Braunschweig.

SMITH, B.J. (1978): The origin and geomorphic implications of cliff foot recesses and tafoni on limestone hamadas in the northwest Sahara. Zeitschrift für Geomorphologie N.F. **22**, 21–43.

STATISTICAL ABSTRACT OF ISRAEL (1950–1982): Central Bureau of Statistics 1–33. Jerusalem.

WILHELMY, H. (1981): Klimamorphologie der Massengesteine. Braunschweig.

Address of author:
Konrad Rögner
Universität Trier
P.O.Box 3825
D-5500 Trier
West Germany

MESURES CONTINUES DES TEMPERATURES DANS LE SOCLE GRANITIQUE EN REGION SOUDANIENNE (FEVRIER 1982–JUIN 1983, OUAGADOUGOU - BURKINA FASO)

M. **Mietton**, Chambéry

Summary

The aim of the experimental temperature-measuring apparatus, sited at Ouagadougou (Burkina Faso, formerly Upper-Volta) is to determine, in a sudanese tropical climate where seasons are highly contrasted, thermal variations in the rock present (granite). In order to do this, five thermo-electric probes are fitted, at depths of -2, -5, -10, -20 and -50 cm, a sixth being situated 0.5 cm above the rock level. These probes are connected to a galvanometric recorder. In the light of the 25.000 items of computer-treated data, temperature variations are first examined for a given depth, following different time scales. Over a year, maximum amplitude is 37°3C at -2 cm. The daily range is 24°7C. Secondly, the range is measured at the same time, but for different depths. The maximum difference with the atmosphere is found at the deepest level in the rock (-50 cm), where there is quasi thermal stability: it is about 12°C. At last, the maximum range inside the rock is higher, in the order of 15°C, between -2 and -20 cm.

This knowledge of thermal variations **in situ** gives a better guidance of experiments carried out in laboratory concerning fatigue of rock surfaces by insolation weathering.

Résumé

Le dispositif expérimental de mesures des températures, situé à Ouagadougou (Burkina Faso, ex Haute-Volta) a pour but de cerner, sous un climat tropical soudanien à saisons contrastées, des variations thermiques dans la roche en place (granite). Pour cela, cinq sondes thermoélectriques sont installées à -2, -5, -10, -20 et -50 cm; une sixième est à $+0.5$ cm au-dessus de la roche. Elles sont reliées à un enregistreur galvanométrique. Les variations de températures à une profondeur donnée, suivant différentes échelles de temps, sont examinées en premier lieu, à la lumière des 25.000 données traitées sur ordinateur. A l'échelle de l'année, l'amplitude maximale est de 37°3C à -2 cm. L'écart quotidien est de 24°7C. En second lieu sont appréhendés les écarts, à un même instant, entre différentes profondeurs.

ISSN 0722-0723
ISBN 3-923381-12-3
©1988 by CATENA VERLAG,
D–3302 Cremlingen-Destedt, W. Germany
3-923381-12-3/88/5011851/US$ 2.00 + 0.25

L'écart maximum avec l'air libre se rencontre dans la roche au niveau le plus profond (−50 cm) assujetti à une quasi stabilité thermique: il est en valeur absolue de 12°C environ. Enfin, l'amplitude maximale à l'intérieur de la roche est supérieure, de l'ordre de 15°C entre −2 et −20 cm.

Cette connaissance des variations thermiques **in situ** peut permettre de mieux orienter les expériences conduites en laboratoire, relatives à la fragilisation des surfaces rocheuses par thermoclastie.

Resumen

El dispositivo experimental para la medición de temperaturas, situado en Ouagadougou (Burkina Faso, ex Alto-Volta) tiene por objeto determinar, bajo un clima tropical sudanés con estaciones contrastadas, las variaciones térmicas que se producen en el granito in situ. Para ello se han instalado 5 sondas termoélectricas a −2, −5, −10, −20 y −50 cm, y una sexta a +0.5 cm sobre la roca. Estas sondas están conectadas a un registrador galvanométrico. Se han obtenido 25000 datos de las variaciones de temperatura a una profundidad determinada, los cuales se han analizado para diferentes escalas de tiempo y en función de la sus variabilidades internas. A lo largo de una año la amplitud máxima se ha registrado a −2 cm y es de 37.3°C y la oscilación diaria ha sido de 24.7°C. La diferencia máxima con la temperatura atmosférica se encuentra en el nivel más profundo (−50 cm), donde hay una casi estabilidad termal alrededor de los 15°C. La variación máxima dentro de la roca se da entre −2 cm y −20 cm y es de 15°C.

Este conocimiento de las variaciones termales in situ puede proporcionar una buena guía para los experimentos llevados a cabo en el laboratorio en relación al desgaste de las superficies rocosas frente a la meteorización por insolación.

1 Introduction

Les mesures de températures **dans la roche en place** sont rares sous toutes les latitudes. Les résultats, dont on dispose, concernent principalement les zones arides (ROTH 1965, PEEL 1974, JÄKEL & DRONIA 1976, SMITH 1977) et subsidiairement les milieux froids (PANCZA 1979). Une meilleure connaissance des variations thermiques **in situ**, à différentes échelles de temps, est pourtant nécessaire pour tenter de cerner, parallèlement aux expériences conduites en laboratoire (COUTARD et al. 1974, JOURNAUX & COUTARD 1974, LAUTRIDOU 1984), leur impact dans la fragilisation des surfaces rocheuses, notamment sous des climats contrastés. Tel est le cas du climat tropical soudanien dans lequel la saison des pluies succède à une période d'échauffement maximum.

La raison majeure de cette lacune tient probablement aux contraintes d'installation du matériel, qui doit être en outre adapté à des contrées climatiquement rudes, parfois difficiles d'accès. Ces difficultés de suivi et aussi de forage, dans un pays comme le Burkina Faso, nous ont conduit à placer notre dispositif expérimental à proximité de Ouagadougou. Cette situation nous a permis en outre de corréler nos mesures de températures aux données pluviométriques ainsi qu'à l'insolation et au rayonnement global relevés dans cette ville.

Quartz	Feldspaths alcalins	Biotite	Hornblende	Apatite	Sphène
26.5%	46.5%	13.7%	11.8%	0.2%	1.3%

Tab. 1: *Composition minéralogique du granite de la carrière de Ouagadougou. (Analyses modales - Institut Dolomieu, Grenoble).*

Tab. 1: *Mineralogical composition of the granite in the Ouagadougou quarry.*

2 Les Caractéristiques de l'expérimentation: site, dispositif et méthode

2.1 Le site

Le dispositif expérimental a donc été installé à la sortie ouest de Ouagadougou, à proximité de la route de Bobo-Dioulasso et d'une carrière désaffectée, où affleure le granite du vieux socle antécambrien. Il s'agit d'un granite à tonalité sodique, dont la composition minéralogique est la suivante (tab.1). La porosité, mesurée au porosimètre à mercure au Centre de Géomorphologie de Caen, est relativement forte (2.6%). La médiane du rayon des pores est de 0.31 microns.

La couleur, relativement blanche dans la masse mais pigmentée par les noyaux de biotite, est gris-clair en surface du fait de l'existence d'une pellicule altérée soumise à la micro-desquamation.

En 1979, ce site avait été retenu pour une première série discontinue de mesures, effectuées au moyen d'un potentiomètre manuel branché sur des thermocouples cuivre-constantan (MIETTON 1980). A partir de 1982, l'utilisation d'un enregistreur galvanométrique[1] et de sondes thermoélectriques à résistance de platine[2] nous a permis d'obtenir des données continues pendant plus d'une année (février 1982 - juin 1983).

2.2 Le dispositif

Cinq sondes thermoélectriques sont placées, après forage au marteau-perforateur, aux profondeurs suivantes: -50, -20, -10, -5 et -2 cm; une sixième est à l'air libre, à 0.5 cm au-dessus de la roche. Elles sont logées dans un support cylindrique en bois, de diamètre équivalent à celui du forage (4 cm), finement usiné longitudinalement pour introduire les fils et transversalement pour les sondes. Ces dernières viennent en contact étroit avec la paroi rocheuse lorsque l'ensemble est enfoncé sous faible pression. La poussière de roche est réintroduite dans les interstices puis un anneau de mastic est placé en surface autour du cylindre, sauf à la verticale des sondes. L'enregistreur est à proximité immédiate, sous une caisse métallique habillée de plaques de polystyrène limitant l'échauffement et arrimée à des pitons à extension. L'énergie électrique est fournie par des piles de 9 volts, assurant une autonomie de l'ordre de 45 jours. Le mécanisme horloger est en revanche remonté toutes les 36 heures seulement, ce qui suppose l'emploi d'un

---

[1] L'ensemble de ces mesures n'aurait pu être réalisé sans l'aide du Service Bioclimatologie de l'INRA à Avignon, qui nous a prêté cet appareil. Nous tenons à remercier son Directeur et tout particulièrement C. BALDY.

[2] Sondes Degussa. Caractéristique: 100 Ω à 0°C.

observateur. L'enregistreur est équipé de six pistes de lecture, interrogées toutes les 30 secondes. Chaque point de mesure fournit une information toutes les 3 minutes. Le tracé de chaque courbe, différenciée par les couleurs des rubans encreurs, apparaît ainsi clairement, de manière quasi continue[3] sur le papier enregistreur. L'établissement d'une courbe d'étalonnage préalable fournit la valeur en degrés Celsius de chaque point.

2.3 La méthode — Choix et limites

- Le choix des profondeurs de mesures résulte d'abord d'une contrainte matérielle. Le diamètre des sondes disponibles (5 mm) interdisait toute implantation dans les premiers millimètres ou même une comparaison détaillée des températures dans la partie superficielle. En outre, du fait d'un gradient thermique très fort dans cet intervalle proche de la surface, la mesure aurait revêtu, avec une telle sonde, un caractère de précision illusoire. Nous avons donc opté pour un premier niveau à −2 cm tout en ayant pleinement conscience que des discontinuités thermiques essentielles peuvent se placer au-dessus! Les quatre autres profondeurs ont été choisies dans un souci de comparaison des températures de la roche en place avec celles des sols, dans lesquels les relevés se font à −5, −10, −20, −50 cm. La confrontation des données, dans le granite et dans une cuirasse ferrugineuse par exemple, peut être instructive, notamment pour la compréhension de leurs réponses spectrales sur les images-satellites. Dans le même esprit, la sixième sonde nous a fourni une référence utile sur la température de l'air au voisinage de la roche.

- La mesure de température se fait à l'extrémité de la sonde, c'est-à-dire à son contact étroit avec la roche. Le bois, choisi comme support, constitue, avec certains plastiques non disponibles, le meilleur isolant possible. Il évite en tout cas tout transfert thermique rapide, de même que l'araldite qui a servi à enrober l'ensemble des fils, dans les parties creusées du cylindre.

- La précision des mesures dépend de l'enregistreur lui-même. Elle a été établie par le constructeur et vérifiée à l'INRA, à Avignon; elle est comprise entre 0.2 et 0.5°C.

L'étalonnage des sondes, réalisé avec un thermomètre à affichage digital est fixé à 0.1°C près. Enfin, la qualité de nos résultats dépend du soin apporté aux opérations de digitalisation lors du traitement des courbes. Finalement, la précision est voisine de ±1°C. Cette marge d'erreur ne réduit pas **la signification des écarts maximums** sur lesquels porte pour l'essentiel cet article et dont l'ordre de grandeur est de 10°C au moins.

3 Les résultats

3.1 Le traitement des données

Les enregistrements, qui peuvent être dépouillés manuellement à l'aide d'une réglette d'étalonnage, ont été traités sur mini-ordinateur (Mini 6)[4] après digitalisation et mémorisation (Tektronic 4052).

---

[3]Distance entre deux points consécutifs: 1 mm.

[4]Le programme Fortran utilisé a été mis au point par M. AYEL du Laboratoire d'Informatique

Mesures des Températures Socle Granitique

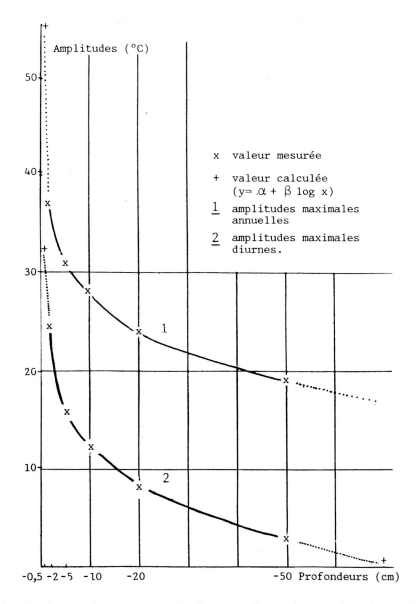

Fig. 1: *Amplitudes thermiques maximales annuelles et diurnes dans le granite en fonction de la profondeur.*

	Profondeurs					
	+0.5 cm	−2 cm	−5 cm	−10 cm	−20 cm	−50 cm
	Températures					
Minimum absolu (7.1.1983)	16°4C	13°7C	17°1C	18°1C	19°7C	22°2C
Maximum absolu (9.5.1983)	48°4C	51°0C	47°8C	46°2C	43°5C	41°3C
Amplitude annuelle	32°0C	37°3C	30°7C	28°1C	23°8C	19°1C

Tab. 2: *Minimums et maximums absolus annuels à différentes profondeurs dans le granite (Ouagadougou).*

Tab. 2: *Annual absolute minimum and maximum values at different depths in granite (Ouagadougou).*

Insolation cumulée	1er Janvier au 18 Avril	1er Mars au 18 Avril
1982	936 heures	385.6 ($\overline{M} = 7.8$)
1983	966.7	426.1 ($\overline{M} = 8.7$)
Ecarts	30.7	40.5

Tab. 3: *Différences de durées d'insolation de Janvier à Avril 1982 et 1983.*

Tab. 3: *Differential durations of insolation between January and April in 1982 and 1983.*

Cette méthode reste longue surtout si l'on retient tous les points sur chaque courbe (480 points par courbe de 24 heures soit 2880 points pour les 6 niveaux de mesures). En fait, la comparaison des résultats de quelques journées, en prenant en compte l'ensemble des points ou bien un point tous les quarts d'heure, nous a montré une grande similitude. On a dès lors retenu la seconde solution. Il faut préciser qu'il ne nous était pas possible, avec le programme dont nous disposions, d'utiliser un pas de temps variable, d'un quart d'heure en général ou de trois

---

Appliquée de l'Université de Savoie. Nous l'en remercions.

minutes au moment des changements de températures les plus rapides.

L'ensemble des informations recueillies entre février 1982 et juin 1983 n'est pas traité ici. Seuls quelques intervalles significatifs ont été retenus: 45 journées, soit 270 courbes, 25.920 points. Les résultats sont obtenus par épisode de 24 heures. Sur chacun de ces tableaux quotidiens, l'espérance mathématique correspond à la moyenne des 96 valeurs saisies tandis que la moyenne est ici égale à la demi-somme du minimum et du maximum. Outre l'écart simple entre ces deux valeurs, "l'écart-max." permet d'appréhender la différence la plus grande entre deux points consécutifs, à une heure qui est précisée. Enfin, l'écart maximum entre chaque niveau et l'air libre (+0.5 cm) d'une part ou bien entre chaque niveau et la profondeur −2 cm d'autre part est spécifié ainsi que sa référence horaire.

Les températures peuvent être examinées en effet de deux manières: dans leurs variations, à une profondeur donnée, suivant diverses échelles de temps et aussi selon leurs écarts, à un même instant, entre différents niveaux.

3.2 Les variations de températures à une même profondeur:

3.2.1 A l'échelle de l'année

Le tab.2 regroupe ci-dessous les minimums et maximums absolus pour chacune des profondeurs et les amplitudes correspondantes. Ces amplitudes **mesurées** $(A = T_x - T_n)^5$ s'ajustent remarquablement bien ($R^2 = 0.99$) suivant une fonction du type $y = \alpha + \beta \log x$ (TOMASSONE et al. 1983), assymptotique en y (fig.1). La question se pose de savoir si l'on peut ajouter à ces valeurs mesurées des valeurs calculées par extrapolation, notamment dans la tranche comprise entre 0 et −2 cm. Cette extrapolation ne se justifierait pas si elle consistait à établir, **à un instant donné**, la température à −0.5 cm par exemple. En revanche, elle est applicable aux **amplitudes**. Les extrêmes (T_x ou T_n) sont enregistrés en effet à **différents instants** aux différentes profondeurs et ils sont évidemment renforcés à un moment quelconque de l'année (ou de la journée) au niveau le plus proche de la surface.

C'est ainsi qu'à −0.5 cm l'amplitude

[5] La précision de ces valeurs est toujours de ±1°C.

maximale annuelle peut être évaluée à 57°C environ. En profondeur, on peut admettre qu'il n'existe plus de variation même annuelle au-delà d'1,80 m à 2 m. On retrouve ainsi le chiffre cité par J. DRESCH pour un granite en zone aride (DRESCH 1966).

Le tab.2 fait apparaître que les valeurs extrêmes sont observées à −2 cm et non pas à l'air libre. Ce classement se vérifie pour toutes les journées analysées, quel que soit le mois. C'est non seulement la surface de la roche mais **toute la tranche d'épaisseur comprise entre 2 et 5 cm qui se réchauffe ou se refroidit davantage que l'air à son contact.**

Ces valeurs extrêmes sont en outre partout enregistrées le même jour: 7 janvier 1983 pour le minimum absolu, 9 mai 1983 pour le maximum absolu. La corrélation est assez bien vérifiée entre ces températures extrêmes et l'insolation ou le rayonnement global, le même jour ou le précédent. Le minimum absolu succède ainsi à l'insolation la plus faible enregistrée après le solstice d'hiver (9.6 heures en moyenne par jour entre le 22 décembre et le 5 janvier; 7.5 heures le 6 janvier; 9.8 heures le 7 janvier). Le maximum absolu correspond à la plus forte insolation d'avril et mai (11,2 heures le 7 mai).

La date des maximums peut être mise en relation, non seulement avec la durée, mais avec la qualité de l'ensoleillement, la verticalité des rayons solaires. Le second maximum observé l'est en effet le 19 avril 1983 (49°2C à −2 cm), au moment du passage du soleil au zénith (20 avril) à Ouagadougou (12°22′N).

D'autres facteurs sont à prendre en compte qui perturbent l'insolation: ce sont bien évidemment les précipitations et les temps couverts qui les accompagnent. A titre d'exemple, le maximum enregistré le 19 avril 1982 atteint 45°6C à −2 cm, valeur inférieure de 3°C à celle du 19 avril 1983. Pourtant, la durée d'insolation est de 11 heures le 19 avril 1982 (rayonnement global: 2556 $j \cdot cm^{-2}$/jour) contre 8.2 heures seulement l'année suivante[6]. Cet écart ne peut être apprécié qu'en fonction des différences d'insolation cumulée le mois précédent, elles-mêmes mises en relation avec l'occurrence des précipitations. Le tab.3, qui permet une comparaison des durées d'insolation durant les quatre premiers mois de 1982 et 1983, montre clairement que c'est en mars-avril qu'il faut rechercher une explication.

La différence d'insolation de 40.5 heures (soit 5 journées environ) est en effet supérieure à celle calculée sur une durée plus longue entre janvier et avril. La cause première en est l'arrivée plus précoce des pluies en 1982: 70.9 mm tombent en 9 averses avant le 19 avril 1982 (dont deux de 37.1 mm et 7.6 mm les 14 et 15 avril) contre 0.1 mm seulement avant le 19 avril 1983. Ces épisodes pluvieux représentent un important déficit calorifique du fait d'une moindre insolation, d'un refroidissement direct sous l'averse et d'un refroidissement différé lié à l'évaporation de petites flaques sur la roche.

Dans le même ordre d'idées, il faut souligner que le maximum, qui pourrait être lié au second passage du soleil au zénith le 21 août, est totalement effacé durant la saison des pluies comprise entre mai et octobre (38°3C à −2 cm le 21 août 1982 par temps couvert). A l'échelle de l'année, la courbe des variations de température est donc nettement dissymétrique; leur croissance est plus rapide que leur abaissement. En 1983, l'amplitude de 37°3C (tab.2) s'inscrit en fait dans un intervalle de 122 jours seulement, entre le 7 janvier et le 9 mai. Nous retrouvons pareille dissymétrie à l'échelle de la journée.

3.2.2 A l'échelle de la journée

Le classement des amplitudes journalières est identique à celui des amplitudes annuelles comme le montre le tab.4. L'écart maximum, proche de 25°C, supérieur de 7°C à celui enregistré en surface, est observé à −2 cm. La décroissance de ces valeurs en profondeur se fait au même rythme que les amplitudes annuelles (fig.1) et l'on peut estimer que les températures journalières sont quasi constantes à −70 cm environ, quels que soient les phénomènes climatiques de surface. Très généralement, la courbe thermique est étale dès −50 cm; nous vérifions ainsi un résultat obtenu, grâce à des mesures discontinues, en 1980 (MIETTON 1980). Les coefficients de corrélation, calculés sur l'ensemble des résultats journaliers, entre les écarts quotidiens en surface d'une part et à chaque profondeur d'autre part sont excellents jusqu'à −10 cm (R = 0.96 à −2 cm; 0.99 à −5 cm; 0.93 à −10 cm), encore significatifs à −20 cm (R = 0.72) mais non significatifs à −50 cm (R = −0.48).

Insolation, rayonnement global et éventuellement précipitations se conjuguent une nouvelle fois pour rendre compte de ces variations sur 24 heures. A −2 cm de profondeur, l'écart de 24°7C correspond à la plus faible insolation (6.8 h. le 9 et le 10 février) enregistrée depuis le début de l'année ce qui creuse le min-

---

[6]Durée d'insolation et rayonnement (BALDY 1985) sont mesurés à la station Ouagadougou-Aérodrome, à 5 kilomètres environ de notre site. Les valeurs du rayonnement global, manifestement sous-estimées en 1983, sont malheureusement inutilisables cette année-là.

	Profondeurs					
	+0.5 cm	−2 cm	−5 cm	−10 cm	−20 cm	−50 cm
Amplitude maximale journalière (date)	Températures 17°7C 19.4.82	24°7C 10.2.82	15°5C 12.10.82	12°4C 12.10.82	8°3C 12.10.82	3°1C 30.5.82

Tab. 4: *Amplitudes journalières maximales dans le granite et à l'air libre.*

Tab. 4: *Maximum daily ranges of temperatures inside granite and in the atmosphere.*

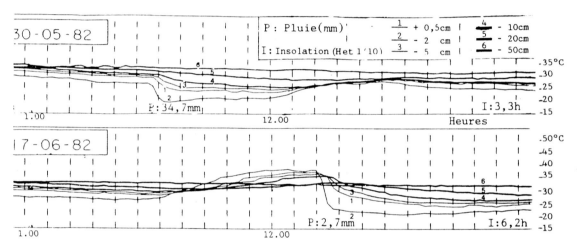

Fig. 2: *Variations thermiques modérées par temps couvert (30-5-1982) ou brutales, sous averses (30-5 et 17-6-1982).*

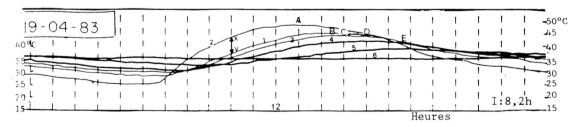

Fig. 3: *Variations thermiques habituelles sous ensoleillement.*

Légende: A, B, C, D et E = heures moyennes des maximums journaliers et écarts-types (σ) sur l'ensemble des journées d'observations

A (−2 cm) = 13h ($\sigma = 0.45$); B (+0.5 cm) = 14h30′ ($\sigma = 0.39$); C (−5 cm) = 15h ($\sigma = 0.25$)
D (−10 cm) = 16h ($\sigma = 0.50$); E (−20 cm) = 17h45′ ($\sigma = 0.56$)
xy = Amplitude maximale, le 19 avril 1983, de 6°5C à 10 heures, entre −2 cm et +0.5 cm.

Date	Pluie (mm)	Amplitude thermique en 15 minutes (à −2 cm)	Amplitude maximale sur une durée ≥ à 15′ (à −2 cm)	Heure pluie	Heure amplitude maximale
20.5.83	18.6	−9°6C	−12°5C en 35′	20h45–20h49 puis 21h30–2h15	22h–22h35
17.6.82	2.7	−9°1C	−14°C en 42′	14h– ?	14h–14h42
20.6.82	18.2	−8°8C	−8°8 en 15′	23h10–0h20	23h15–23h30
12.10.82	15.0	−7°5C	−11°C en 47′	18h10–21h10 puis 21h25–1h25	18h18–19h05
2.8.82	20.9	−5°1C	−12°5C en 71′	8h02–9h20 puis 12h10–13h15	12h06–13h17
27.5.82	27.0	−5°C	−7°C en 28′	2h35–4h15 puis 5h48–7h45	2h45–3h13
30.5.82	34.7	−4°5C	−7°5C en 33′	6h37–11h37	6h42–7h15
4.6.82	2.8	−4°1C	−6°C en 27′	18h55–19h40	18h51–19h18
12.6.82	0.6	−3°6C	−5°5C en 24′	20h35–22h10	20h36–21h
25.6.82	4.3	−3°2C	−4°5C en 24′	23h06– ?	23h06–23h30
18.7.82	28.6	+4°3C	−5°5C en 38′	0h55–5h25	−5°5C entre 0h55 et 5h25 +4°3C entre 9h30 et 9h45
17.2.82	0.1	+1°9C	pluie sans effet	20h30–21h05	7h30–7h45 échauffement matinal
19.4.83	traces	+2°1C	pluie sans effet	16h03	6h45–7h échauffement matinal
31.10.82	traces	+2°3C	pluie sans effet	15h10	8h15–8h30 échauffement matinal

Tab. 5: *Variations de température à la surface de la roche (−2 cm) sous averses.*

Tab. 5: *Temperature patterns at two centimeters below the surface (−2 cm) under sudden showers.*

imum (17°2C) tandis que le maximum (41°9C) est déjà bien supérieur à celui enregistré le 7 janvier (30°8C), date du minimum annuel absolu (13°7C).

Autre exemple, les amplitudes quotidiennes extrêmes du 12 octobre 1982 (22°9C à −2 cm) s'expliquent par un très fort rayonnement global de 2468 joules · cm^{-2}/jour (maximum absolu: 2647 j. · cm^{-2}/jour) et une pluie de 15 mm ne limitant pas l'insolation (10.2 h.) puisque commençant à 18h10 pour s'interrompre momentanément à 21h10 puis définitivement à 1h25. En revanche, l'averse du 30 mai 1982 (34.7 mm), débutant beaucoup plus tôt, à 6h37, et s'accompagnant de très faibles rayonnement (785 j · cm^{-2}/jour) et insolation (3.3 h.), se fait sentir dans la même journée jusqu'à −50 cm (fig.2).

Outre le problème de l'épaisseur de la roche pouvant être soumise aux influences climatiques externes, c'est celui de leur vitesse de propagation qui peut être posé. L'interprétation s'appuie dès lors sur les comparaisons des heures d'occurrence des maximums ou des minimums entre différents niveaux.

Nos mesures corroborent des faits établis depuis longtemps dans des sols meubles (GEIGER 1966, SELTZER 1935) et précisent les données plus rares dans la roche en place:

- la température maximale est atteinte en début d'après-midi, après le passage du soleil au zénith, avec un décalage qui augmente en profondeur, de l'ordre de 12 heures à −50 cm. A 10 cm de profondeur, la température continue à s'abaisser pendant une heure à une heure trente après que la surface ait commencé à se réchauffer (PEEL 1974);
- le refroidissement progressif nocturne atteint son maximum, par ciel clair, quelques minutes ou une heure après le lever du jour. Le décalage n'est net qu'à −50 cm où le minimum est enregistré vers 14h-14h30, c'est-à-dire au moment du maximum à l'air libre (MIETTON 1980). Les courbes thermométriques présentent ainsi un tracé nettement dissymétrique, étalé à droite (fig.3), avec un temps d'échauffement d'environ huit heures, deux fois plus rapide que le refroidissement. Ce dernier est accéléré jusque vers 18h30 avant de se ralentir durant la soirée et la nuit, ce que R.F. PEEL (1974) avait déjà constaté.

Dans le détail, ces vitesses de propagation n'apparaissent pas constantes tout au long du profil mais maximales entre 5 et 20 cm. Cette différence peut venir d'une meilleure cohésion des grains, les micro-fissures étant probablement plus nombreuses en surface. La compacité du granite assurerait ainsi davantage de conductibilité thermique.

On peut remarquer en outre que certaines journées à ciel couvert et très faible insolation peuvent être caractérisées par une quasi stabilité thermique à tous les niveaux. Ce phénomène peut s'accompagner d'une totale inversion de température, le niveau −50 cm restant le plus chaud durant toute la journée suite à une averse durable (fig.2, 30 mai 1982).

3.2.3 A l'échelle du quart d'heure

Les tab.5 et 6 regroupent les informations relatives aux écarts maximums, enregistrés généralement en 15 minutes. Les enseignements sont multiples:

Dates (pluies en mm)		+0.5 cm	−2 cm	Profondeurs −5 cm	−10 cm	−20 cm	−50 cm
20.5.83 (18.6 mm)	A	−2°5C	−9°6C	−1°8C	+1°1C	+0°5C	−0°5C
	H	22	21.5	22.7	10	14.2	10.5
17.6.82 (2.7 mm)	A	−4°C	−9°1C	−2°1C	−1°5C	−0°4C	−0°5C
	H	14.5	14.2	14.5	14.7	0.5	14.0
20.6.82 (18.2 mm)	A	−2°4C	−8°8C	−1°3C	−1°1C	−0°5C	−0°4C
	H	23.5	23.2	23.5	0.0	20.2	4.5
12.10.82 (15.0 mm)	A	−2°6C	−7°5C	−2°C	−1°1C	0°8C	0°4C
	H	18.7	18.7	19.0	19.2	9.0	16.0
2.08.82 (20.9 mm)	A	−2°9C	−5°1C	?	−1°1C	−0°7C	0°6C
	H	12.2	12.2		13.0	14.2	11.0
27.5.82 (27.0 mm)	A	−2°C	−5°C	−1°2C	−1°1C	−1°2C	−0°5C
	H	3	2.7	3.2	18.2	1.7	14.7
30.5.82 (34.7 mm)	A	−1°9C	−4°5C	−1°6C	−0°8C	−0°5C	0°8C
	H	6.7	6.7	7.0	7.2	1.0	3.5
4.6.82 (2.8 mm)	A	−1°7C	−4°1C	−1°3C	−1°1C	+1°1C	−0°5C
	H	18.5	18.7	19.0	17.7	15.2	22.2
12.6.82 (0.6 mm)	A	−1°4C	−3°6C	−1°1C	−0°9C	−0°9C	+0°4C
	H	20.7	20.2	20.7	21.2	20.2	23.0
25.6.82 (4.3 mm)	A	−1°4C	−3°2C	−1°2C	+1°2C	+0°7C	−0°4C
	H	23.2	23.2	17.7	8.7	10.2	8.7
18.7.82 (28.6 mm)	A	+1°6C	+4°3C	−1°7C	+1°C	+0°6C	−0°4C
	H	9.7	9.5	1.5	10.5	11.7	20.7
17.2.82 (0.1 mm)	A	+1°5C	+1°9C	−1°1C	−1°2C	+0°5C	−0°5C
	H	10.2	7.5	19.2	19.0	14.7	6.5

Tab. 6: *Amplitudes (A) maximales enregistrées en 15 minutes et horaires (H) correspondants (les quarts d'heure sont notés .2, .5 ou .7).*

Tab. 6: *Maximum ranges of temperatures (A) recorded in 15 minutes and corresponding times (H) (The quarters of an hour are noted. .2, .5 and .7.)*

- Les variations de températures les plus accusées sont négatives et consécutives à une averse (tab.5). Elles sont nettement plus sensibles à la surface de la roche (−2 cm) qu'à l'air libre; le rapport des écarts étant généralement voisin de 2.5 mais pouvant atteindre 3.8 (20 mai 1983 et 17 juin 1982). La chute de température est également plus immédiate mais se prolonge moins durablement.

- L'abaissement maximal observé à −2 cm est de 9°6C en 15 minutes (20 mai 1983) et de 14°C en 42 minutes (17 juin 1982, fig.2). De manière plus instantanée, la chute de température est de 5°C en 6 minutes le 20 mai 1983[7].

- Ces écarts se font sentir jusqu'à −10 cm le plus souvent, quelquefois 5 cm (20 mai 1983, 27 mai 1982, 4 juin 1982), exceptionnellement 20 centimètres (tab.6). Cette détermination ne tient pas seulement compte du caractère négatif de l'écart tel qu'il peut être signalé dans chacun des tableaux unitaires mais de l'horaire correspondant qui rend vraisemblable ou non la corrélation avec l'averse. En moyenne, l'écart n'est plus que de 1°5C à 5 cm (maximum observé: −2°1C) et de 1°C à −10 cm (maximum observé: −1°5C).

- L'amplitude thermique n'est pas corrélée de manière simple avec la hauteur de pluie. Deux petites averses identiques (2.8 mm le 4 juin et 2.7 mm le 17 juin 1982), à moins de 15 jours d'intervalle, déterminent des baisses de températures très différentes de 4°1C et 9°1C. L'orage majeur du 30 mai 1982 (34.7 mm) n'a pas davantage d'impact (−4°5C).

Seules les traces du 17 janvier, 31 octobre 1982 et 19 avril 1983 n'ont pas ou peu d'effet, dans tous les cas inférieur à l'échauffement matinal de l'ordre de 2°C en 15 minutes (écart positif maximum observé: +3°1C le 11 novembre 1982 à 7h30). La pluie de 0.6 mm (12 juin 1982) à laquelle est liée une baisse de 3°6C peut être considérée **a priori** comme une valeur limite; elle doit être considérée en fait avec prudence dans la mesure où les relevés pluviométriques sont faits à quelque distance et l'on sait le caractère souvent ponctuel des orages. Dans tous les cas, il semble bien cependant qu'il suffise de quelques millimètres de pluie pour déclencher une baisse sensible de température à la surface de la roche. Ceci revient à dire que **la fréquence des chocs thermiques en est plus grande**.

- Ces chocs thermiques correspondent très largement au passage de lignes de grains s'abattant en début de saison des pluies (mai-juin) ou secondairement à la fin (pluie du 10 octobre 1982), sur une dalle rocheuse préalablement surchauffée. Ainsi, l'averse du 20 mai 1983, pour laquelle on enregistre des amplitudes maximales, n'est précédée que d'une seule pluie (15.6 mm le 17 mai) en ce début d'hivernage 1983.

A l'échelle de la journée, l'heure d'occurrence joue aussi un rôle capital. La petite pluie du 17 juin 1982, remarquable par son efficacité (tab.5), survient à 14h, au moment

---

[7] Les déterminations horaires ont été faites, dans les cas les plus intéressants, non pas sur les courbes obtenues après digitalisation (fig.2 et 3) mais par un examen détaillé des diagrammes d'origine.

de l'échauffement maximal. Il en est de même le 2 août. Les autres averses efficaces se placent en fin d'après-midi ou dans la soirée; au contraire, les fortes pluies à influence limitée tombent au coeur de la nuit (27 mai 1982) ou tôt le matin (30 mai 1982).

- Les décalages horaires dans l'amorce des chutes de températures aux différentes profondeurs permettent une nouvelle fois d'apprécier la vitesse de propagation de l'onde de refroidissement (tab.6). Les moyennes des écarts sont de trente minutes environ entre les niveaux −2 et −5 cm puis 20 minutes entre −5 et −10 cm, soit des vitesses de 6 cm/h et 15 cm/h. On vérifie ainsi l'accélération de la propagation de l'onde thermale en profondeur (dans un rapport proche de 2 comme précédemment). Mais cette vitesse est, dans l'absolu, trois ou quatre fois plus élevée que dans le cas d'un simple refroidissement nocturne.

3.3 Les écarts de températures entre niveaux au même instant

3.3.1 Les contrastes thermiques roche — air libre

Les écarts principaux avec l'air libre, occasionnellement soumis à des variations climatiques rapides, sont à rechercher généralement au niveau le plus profond de la roche (−50 cm), assujetti à une quasi inertie thermique. Ce contraste peut être positif ou négatif; nous nous sommes bornés à classer les cinq premières valeurs dans chacun des deux cas (tab.7). On constate que l'amplitude roche (−50 cm) — air libre est au plus égal à 12°C.

Les écarts sont positifs, autrement dit l'atmosphère est plus chaude que le granite, en début d'après-midi (14–15 h) de saison sèche ou mieux de fin de saison fraîche (4 valeurs sur 6 relevées en février; 2 en avril). En revanche, le coeur de la roche peut être plus chaud, au moment des averses bien sûr ou, plus précisément, à la fin de l'averse. La comparaison des tab.5 (heure de la pluie) et 6 (heure de l'écart) est très parlante à cet égard.

Toutefois, dans le cas de précipitations, on sait que la surface de la roche se refroidit davantage que l'air et l'on n'est donc pas surpris de retrouver des différences de l'ordre de +10°C entre l'atmosphère et le niveau −2 (10°9C le 20 mai, 9°7C et 9°2C les 17 et 20 juin). A n'en pas douter, on doit retrouver des contrastes majeurs, durant ces journées, entre les points −2 et −50 cm.

3.3.2 Les contrastes thermiques à l'intérieur de la roche

L'examen complet des 45 tableaux unitaires permet de limiter la comparaison entre le niveau −2 cm et les niveaux −20 et −50 cm. Les valeurs absolues les plus élevées se concentrent en effet dans ces deux intervalles. Elles sont classées dans le tab.8.

La surface de la roche étant plus sensible que l'air aux influences climatiques externes, il est logique de trouver des écarts, entre −2 et −50 cm, supérieurs aux précédents, de l'ordre de 14°C à 16°C désormais. Ces effets identiques mais amplifiés sont liés aux mêmes causes: averses de mai et juin pour les écarts négatifs, échauffement de surface en début d'après-midi (13h contre 14–15 h. à l'air libre) et en saison sèche pour les écarts positifs. Ces derniers

	$T_N > T_V$			$T_N < T_V$		
	(+0.5)	(−50)		(+0.5)	(−50)	
	Ecart	Date	Heure	Ecart	Date	Heure
1	+12°C	19.4.82	20	−9°7C	30.5.82	10.7
2	+11°3C	11.2.82	14.2	−8°9C	18.7.82	6.0
3	+11°C	10.2.82	14.7	−8°4C	17.6.82	19.7
4	+10°8C	17.2.82	14.7	−8°2C	27.5.82	7.2
5	+10°5C	14.2.82	15.2	−7°5C	20.6.82	2.7
		19.4.83	13.7			

Tab. 7: *Contrastes thermiques maximums entre l'air libre (T_N : +0.5 cm) et la roche (T_V : −50 cm) à un instant donné.*

Tab. 7: *Maximum ranges of temperature between the atmosphere (T_N : +0.5 cm) and the rock (T_V : −50 cm) at a given time.*

Contrastes thermiques entre −2 et −20 cm				Contrastes thermiques entre −2 et −50 cm			
Ecarts positifs ($T -2 > T -20$)		Ecarts négatifs ($T -2 < T -20$)		Ecarts positifs ($T -2 > T -50$)		Ecarts négatifs ($T -2 < T -50$)	
A	Date et heure	A	Date et heure	A	Date et heure	A	Date et heure
+15°6C	11.2.82 (12.5)	−15°3C	20.5.83	+15°5C	11.2.82 (13.2)	−14°C	30.5.82 (7.7)
+15°4C	10.2.82 (13.0)	−12°5C	30.5.82 (7.7)	+15°1C	19.4.82 (14.5)	−12°3C	27.5.82 (3.5)
+14°1C	19.4.82 (14.5)	−12°5C	20.6.82 (23.7)	+15°C	10.2.82 (13.0)	−11°9C	20.6.82 (23.7)
+13°9C	17.2.82 (11.7)	−11°9C	27.5.82 (3.5)	+14°9C	17.2.82 (13.0)	−11°8C	12.10.82 (1.0)
+13°7C	11.11.82 (11.2)	−11°1C	17.6.82 (16.0)	+14°C	14.2.82 (13.7)	−11°5C	17.6.82 (18.2)

Tab. 8: *Contrastes thermiques maximums (A) à l'intérieur de la roche à un instant donné.*

Tab. 8: *Maximum ranges of temperature (A) inside the rock at a given time.*

sont en valeur absolue plus élevés que les refroidissements sous averses. S'il existe une fatigue de la roche pouvant être liée à des contrastes thermiques entre niveaux, **la fréquence de ce type de contraintes apparaît plus grande qu'on ne l'imaginerait a priori et pas seulement liée à des phénomènes exceptionnels.**

Enfin, il est notable que les amplitudes maximales, parmi toutes celles enregistrées, se placent entre -2 et -20 cm. Le maximum maximorum observé est de 15°6C le 11 février 1982, à 12h30 (tab.8). Cette valeur peut être mise en rapport avec l'écart extrême de 24°C enregistré dans le désert mojave par ROTH (1964), le 10 juillet 1962 à 14h30 environ, entre les niveaux -3.8 et -16.5 cm, durant une période d'observation de 8 mois.

4 Conclusion

Les variations de températures dans la roche en place, sous climat soudanien, sont désormais mieux connues, grâce à ces mesures réalisées en continu et sur une longue durée. Elles pourraient être évidemment précisées par l'emploi de sondes plus petites et placées plus près de la surface.

D'ores et déjà, les écarts de températures, entre niveaux ou bien à une même profondeur, apparaissent de manière significative et surtout répétée, y compris à l'échelle d'une journée commune, sans accident climatique. Toutefois les chocs thermiques les plus brutaux font suite aux averses, même de faible quantité, tombant sur une roche surchauffée. Le refroidissement a plusieurs origines: la perte de rayonnement direct, l'impact de la pluie, l'évaporation des flaques ou des gouttes d'eau qui subsistent à la surface mais aussi une évaporation dans les interstices ou les fissures de la roche. Cette évaporation plus profonde pourrait être plus efficace dans les processus de fragmentation du matériel.

Depuis le travaux de BLACKWELDER (1933) et GRIGGS (1936), de nombreux auteurs continuent de s'opposer sur l'efficacité réelle de ces variations thermiques (OLLIER 1984). La poursuite des expérimentations de laboratoires apparaît très souhaitable afin de mieux cerner tous les paramètres en jeu. Des essais pourraient être repris en imposant à des échantillons de grande taille des amplitudes constatées sur le terrain, accompagnées de phénomènes de vaporisation brutale.

References

BALDY, C. (1985): Contribution à l'étude des applications de la bioclimatologie végétale à l'agrométéorologie des zones arides et semi-arides en climats méditerranéen et tropical. Thèse Marseille. 213 pp., 85 fig., 26 tabl., 8 cartes.

BLACKWELDER, E. (1933): The insolation hypothesis of rock weathering. American Journal of Science, 5e série, vol. **26**, no. **152**, 97–113.

BRUNET, Y. (1984): Modélisation des échanges sol nu-atmosphère. Essai de validation locale et influence de la variabilité spatiale du sol. Thèse Grenoble. INPG, 189 p.

COUTARD, J.P., BENOIST, J.P. & GUILLEMET, G. (1974): Pénétration de la chaleur dans des échantillons de calcaire et de granite. Bulletin du Centre de Géomorphologie du CNRS, Caen, No. **18**, 21–29.

DRESCH, J. (1966): La zone aride. In: Géographie générale. Encyclopédie de la Pléiade. Gallimard. Paris, 730–739.

GEIGER, R. (1966): The climate near the ground. 4th Ed. Harvard Press, Cambridge, Mass., 611 p.

GRIGGS, D.T. (1936): The factor of fatigue in rock exfoliation. Journal of Geology, **44**, 783–796.

JÄKEL, D. & DRONIA, H. (1976): Ergebnisse von Boden und gesteinstemperaturmessungen in der Sahara. Berliner Geographische Abhandlungen, Heft **24**, 55–64.

JOURNAUX, A. & COUTARD, J.P. (1974): Expérience de thermoclastie sur les roches siliceuses. Bulletin du Centre de Géomorphologie du CNRS, Caen, No. **18**, 7–20.

LAUTRIDOU, J.P. (1984): Fabrication expérimentale de gélifracts. Bulletin du Centre de Géomorphologie du CNRS, Caen, No. **27**, 15–38.

MIETTON, M. (1980): Recherches géomorphologiques au Sud de la Haute-Volta. La dynamique actuelle dans la région de Pô-Tiébélé. Thèse 3ème cycle. Grenoble, 235 p.

MIETTON, M. (1988): Dynamique de l'interface lithosphère-atmosphère au Burkina Faso. Contribution géomorphologique à l'étude de l'érosion en zone tropicale de Savane. Thèse de Doctorat d'Etat. Grenoble. 497 p. 107 fig. 94 tab. 227 annexes.

OLLIER, C.D. (1984): Weathering. Second edition. Geomorphology Texts No. **2**, 270 p. Longman, London.

PANCZA, A. (1979): Contribution à l'étude des formations périglaciaires dans le Jura. Bulletin de la Société Neufchateloise de Géographie, No. **24**, 171 p.

PEEL, R.F. (1974): Insolation weathering. Some measurements of diurnal temperature changes in exposed rocks in the Tibesti region, central Sahara. Zeitschrift für Geomorphologie, Suppl. **21**, 19–28.

ROTH, S.E. (1965): Temperature and water content as factors in desert weathering. Journal of Geology, **73**, 454–468.

SELTZER, P. (1935): Etudes micrométéorologiques en Alsace. Doctorat Sciences physiques. Université de Strasbourg. Série E, No. **49**, 57 p.

SMITH, B.J. (1977): Rock temperature measurements from the northwest Sahara and their implications for rock weathering. CATENA, vol. **4**, 41–63.

TOMASSONE, R., LESQUOY E. & MILLIER, C. (1983): La régression, nouveaux regards sur une ancienne méthode statistique. INRA. Actualités scientifiques et agronomiques **13**, 180 p., Masson, Paris.

Address of author:
M. Mietton
Département de Géographie — Université de Savoie
Boîte Postale 1104
73011 Chambéry, France

F. Ahnert, H. Rohdenburg & A. Semmel:

BEITRÄGE ZUR GEOMORPHOLOGIE DER TROPEN (OSTAFRIKA, BRASILIEN, ZENTRAL- UND WESTAFRIKA) CONTRIBUTIONS TO TROPICAL GEOMORPHOLOGY

CATENA SUPPLEMENT 2, 1982
Price: DM 120,–
ISSN 0722–0723 / ISBN 3–923381–01–8

F. AHNERT
UNTERSUCHUNGEN ÜBER DAS MORPHOKLIMA
UND DIE MORPHOLOGIE DES
INSELBERGGEBIETES VON MACHAKOS, KENIA

(INVESTIGATIONS ON THE MORPHOCLIMATE
AND ON THE MORPHOLOGY OF THE
INSELBERG REGION OF MACHAKOS, KENIA)

S. 1–72

H. ROHDENBURG
GEOMORPHOLOGISCH–BODENSTRATIGRAPHISCHER
VERGLEICH ZWISCHEN DEM
NORDOSTBRASILIANISCHEN TROCKENGEBIET
UND IMMERFEUCHT–TROPISCHEN GEBIETEN
SÜDBRASILIENS

MIT AUSFÜHRUNGEN ZUM PROBLEMKREIS DER
PEDIPLAIN–PEDIMENT–TERRASSENTREPPEN

S. 73–122

A. SEMMEL
CATENEN DER FEUCHTEN TROPEN
UND FRAGEN IHRER GEOMORPHOLOGISCHEN
DEUTUNG

S. 123–140

SLOPE EVOLUTION
BY MASS MOVEMENTS AND SURFACE WASH
(VALLS D'ALCOI, ALICANTE, SPAIN)

N. La Roca Cervigón & A. Calvo-Cases, València

Summary

This article deals with the evolution of the gully slopes of the Alcoi valleys (SE Spain), dug out in soft rock. The interpretation of the aerial photograph and on-site observation has made it possible to make a cartography in side a small but representative area, several types of mass movements:

a) old slides,

b) contemporary mudslides.

In order to come closer to the failure conditions and to the mechanism of the second type, which is currently the one most frequently ocurring on these gully slopes, a morphological and morphometric description of one of the mudslides that took place in 1982 has been carried out.

Some information on the soil characteristics which has come to light during laboratory analysis is also presented.

Resumen

Este artículo trata de la evolución de las laderas de las cárcavas de los valles de Alcoi (SE español), excavadas en rocas blandas. La interpretación de la foto aérea y la observación directa sobre el terreno han permitido cartografíar, en un área pequeña pero representativa, distintos tipos de movimientos de masa:

a) antiguos slides y

b) mudslides actuales.

Con el fin de aproximarnos a las condiciones de rotura y al mecanismo del segundo tipo, actualmente el mas frecuente sobre las laderas de estas cárcavas, se ha realizado una descripción morfológica y morfométrica de uno de los mudslides ocurridos en 1982.

Se ofrece asimismo información sobre las propiedades del suelo obtenidas del análisis en el laboratorio.

1 The Study Area

The study area is (fig.1) in one of the structural depressions at the NE end of the Betic Mountains. The geological substratum consists of Miocene marine white marl series, nearly 1000 m deep, locally covered by conglomerate deposits of the Pliocene-Quarternary. These materials are dissected by a dense network of gullies, all of them tributaries of the Alcoi or Serpis River.

ISSN 0722-0723
ISBN 3-923381-12-3
©1988 by CATENA VERLAG,
D–3302 Cremlingen-Destedt, W. Germany
3-923381-12-3/88/5011851/US$ 2.00 + 0.25

The climate, with annual rainfall values between 470 mm and 950 mm, is temperate Mediterranean with warm thermal and humid Mediterranean regimes depending on location. At Cocentaina station (altitude of about 400 m above sea level) for example, a precipitation of 155 mm in 24 hours has a recurrence period of ten years and a continuous rainfall, or storm of 2–6 days and 290 mm has the same frequency.

Temperatures also show important contrasts. The annual mean temperature is 14.5°C and a maximum mean value of 24.6°C occurs in July or August and a minimum mean value of 8°C in January. The mean of the maxima of July is 31°C. The most frequent daily amplitude in a year lies between 9–15°C (46%); it is greater in summer: between 12–18°C (56.8%). The average number of frost days in a year is 26 and frost occurs during 5 months from November to March.

The vegetation cover is poor. The climax vegetation (*Quercetonion rotundifoliae*), a forest of *Quercus rotundifolia* has practically disappeared. Today the forest is generally formed by Aleppo pines (*Pinus halepensis*) and some maritime pines (*Pinus pinaster*) with shrubs. When degradation is at its most severe, the trees disappear and the succession is: maquia, garriga and finally a xerophilous steppe (*Brachypodium* sp., *Stipa* sp., etc.).

Another factor affecting the morphogenesis of this area is seismic movements. Since 880 BC thirty three earthquakes have, for a certainty, affected this region, and probably quite a few others. Among them we very specially want to stress six, which reached or surpassed the grade VI in the M.S.K. scale (MEZCUA & MARTINEZ 1983). According to these authors cracks and mass movements began to appear from grade VI upwards.

2 Slope Morphology

This paper concentrates on the "les Carxofes" gully, one of a series of tributaries to the Alcoi River, which has developed on a system of hanging pediments lying at heights between 600 and 400 m a.s.l. These gullies cut into the underlying marls to a depth of about 50–40 m, near their mouth. Their slopes evolve rapidly through mass wasting and wash processes, probably in a similar manner to the formation of badlands.

The starting point of the Carxofes gully slopes is the undulating surface corresponding to the pediments, which near the Alcoi River pass into an old terrace. Fig.2 shows the measurements, on the 1:5000 scale map (fig.1), of some slope profiles which reveals their mean form (counterline equidistance of five metres). The slope morphology is very irregular. The transition towards the upper pediment-terrace frequently takes on the shape of a cliff (profiles A, E, F, G). These data, along with the results of the interpretation of the aerial photograph (approx. scale 1:18000) and the on-site inspection, indicate that the mass movements are of greatest importance amongst the processes that influence the slope form. In fig.1 we have ditinguished two types of mass movement:

a) slides,

b) mudslides/flow.

Among the slides at least two generations can be identified, the most recent one with movements that affect a lesser quantity of material, but the ages of those two generations are unknown. However, the fact that the slides that belong to the oldest series are terraced points to the existence of fossil slides.

Fig. 1: *Morphological scheme of the Carxofes gully. 1) Cliff; 2) Deep-seated slides; 3) Mudslides without a flow tongue; 4) Mudslide/flow 1982; 5) Mudslide/flow 1986; 6) Gully channel; A, B, C, D, E, F, G and H: location of slope profiles of fig.2.*

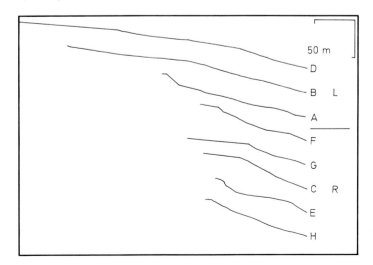

Fig. 2: *Slope profiles along the valley (see location in fig.1).*

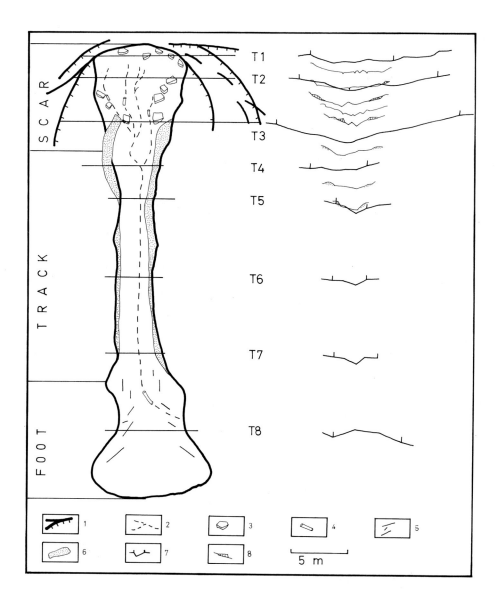

Fig. 3: *Plan and cross sections of the mudslide. 1) Scarp and cracks; 2) Rills; 3) Shallow slides; 4) Pipes; 5) Crossed cracks at the foot; 6) Mud accumulation, levées; 7) Cross profiles; 8) Shallow slides in cross profiles.*

Among the mudslides considered here are those that took place in the wake of the 1982 and 1986 rains.

The mudslides originate in tensional cracks, which open up as a result of strong breaks of slope and therefore the scars left by the mudslides are frequently present in the cliff.

3 The Mudslide

The mass movement that we studied occured between the months of October and November in 1982. Its failure was provoked by continuous and heavy autumn rains after a long dry summer. In the four months from July to September only 35.5 mm rain were recorded in the neighbouring town of Almudaina (586 m altitude). The rainy period began on October 18th and during five consecutive days a total of 153.4 mm rain fell. After seven days without rain another storm occurred on the 30th of October and three following days, which amounted to a total of 346.6 mm rain distributed as follows: 91, 124.2, 130.5. Although the exact time when the mass movement began, is not known it is believed to have occurred during the last period of heavy rainfall, after reactivation of piezometric levels and subsurface pressures.

The mass movement lies in the upper part of a slope concave in cross section. It starts from a small step 10 m from the crest and spreads down 50 m, ending at a strong break of slope. Its maximum width is 10–11 m, at the foot. The volume of displaced material was 32.51 m^3 with a weight of 38.4 metric tones (density used 1.2 gr/cm^3). There may have been some overestimation because we considered the original surface as being flat.

The slope on which this mass movement occurs has at present a mean gradient of 28°, with 34° on the upper third and 26° on the remainder. Its form is very irregular due to old cultural terraces and older movements.

The form of the mudslide consists of three clear parts: source area or scar, channel or track and foot or accumulation area (fig.3).

a) The source area is about 50 m^2, with a mean gradient of 37°. On the slide surface, which is flat or smoothly concave, polished and grooved, there are some remains of earth blocks, mainly near the scarps, ready to slide downslope. They have dimensions of about 40 cm deep, 50 cm wide and 70 cm long. Surrounding the scarp there is a series of concentric cracks, 40 cm deep which is approximately the regolith thickness on this part of the slope.

b) The track: Its longitudinal profile, with parts at 43° and other parts with a gradient of only 19°, reflects the previous topography.

Cross-section form is masked by remains of the transported materials consisting of marly mud that has lost its original structure. This material overflows the channel banks and lies on the untouched slope forming levees 20 cm thick similar to those described by BRUNSDEN (1984), JOHNSON (1984) and COLIN ROUSE (1984).

Along the track there are several signs that surface wash occurred after mud transportation in a dense flow: vegetation has been combed downslope and the cross-section is V-shaped (fig.3). Wash activity has been verified by placing sediment traps (YOUNG 1972), three years after the mudslide occurrence. In two months (October–November 1985) the trap at the lower part of the track was filled with sediments and overflowed.

c) The foot or accumulation area is cone-shaped, some 75 cm thick. Longitudinal and cross profiles are convex. The surface is uneven, with small heaps and near the apex there is a system of crossed cracks, oblique to flow direction, related to the sharp decrease in velocity (DELAPUE et al. 1964).

The material has a soft spongelike consistency when wet. However in the dry season it hardens.

Amid the processes shaping the mudslide tongue we stress the formation of pipes that when they collapse produce rills 20–30 cm in width.

Fig.3 shows two series of cross profiles of the mudslide, which emphasize the above mentioned distribution, evacuation and accumulation zones. The cross profile series with more detailed measurements reflect the slight variations that the subsequent wash has caused on the scar. It is reshaped by rills and recently we have observed there subsurface wash (pipe development) becoming increasingly important.

4 Morphometry and Mechanics

To illuminate the genesis and mechanics of this type of process we attempted (in addition to the observation of micromorphology) to apply CROZIER's (1973) morphometric indices on the mudslide form under study. The results are:
Classification 0.01
Dilation 1.3
Displacement 72%
Tenuity 3.5
Flowage 105%

From CROZIER's (1973) values we find that the dominant features are those of "fluids flows". The ratio between maximum width and total length is the main factor that determines this classification. In the upper part of the source area the movement form or type is that of a shallow slide. This conclusion consideration is based on micromorphological analysis. This movement is therefore defined as "mudslide" according to BRUNSDEN (1984).

The values obtained for the indices of classification, dilation and tenuity are below the mean values obtained by CROZIER, because of the long narrow shape of the mudslide. On the one hand, this can be related to the high gradient (about 30°); on the other, the smaller basal accumulation in comparison to the source area shows the importance of some loss either by channel storage or by surface wash. The flowage index, which is higher than the maximum values obtained by CROZIER (1973) shows that the flowage amount was very high. This suggests, together with the track and foot morphology, that the mudslide passes into a mudflow.

In areas like the one on which the movement lies, concave topography helps subsurface water to become concentrated. The rapid rise in piezometric levels increased downslope.

5 The Material

In the slope sections without recent mass movements the surface consists of a layer 40–50 cm thick of disturbed material. Although the $CaCO_3$ content is very high (50–70%), which reduces volumetric change, wetting/drying cycles and subsurface wash activity through piping, make this surface layer very cracked, dry and loose in summer. The reaction of this utterly dry material when subjected to fast saturation is to slake. This, together with the consequences of sliding, may largely explain the transition from a slide to a flow.

Soil samples were taken from the scar on the 14. November 1986 and analysed in the laboratory. The results are as follows:

Atterberg limits: liquid limit 58; plastic limit 24; plasticity index 34. Therefore material is a high-plasticity soil: USCS (Casagrande): CH.

The specific gravity of the solids (Gs) is: 2.7 gr/cm^3.

On that day the water contents at different depths in a probe carried out on the scar, on the more compact underlying material, i.e. under the slide surface, were as follows: surface crust: 6.31%; at 15 cm depth: 20.92%; at 45 cm depth: 25.85% and at 60 cm depth: 25.38%.

Further down a block sample was extracted for the unconfined compression test. The water content, w, inside the block was 27.94%, i.e. the soil was to practical purposes at its plastic limit.

This material has undergone a certain amount of weathering and shows a series of tiny cracks where it breaks apart during the test, which give the following result: qc = 0.35 Kg/cm^2, with a bulk density (wet density) of 1.88 gr/cm^3 and a dry density of 1.47 gr/cm^3. The void ratio, e, is e = 0.836, with a degree of saturation of 89.9%. It is important to point out that the void ratio corresponding to the LL should be 1.56 and if the soil should not expand the water content would reach the 30.9% level when saturation occurred.

The unconfined compression test gives an undrained shear strength ($\phi = 0$ analysis) of 1.75 Tm/m^2. Assuming an infinite slope (GRAHAM 1984) with an inclination of 28°, the critical depth for mobilizing shear strength is approximately 2 m; plastic analysis with Coulomb model (JOHNSON 1984) gives 1.98 m.

No information is available concerning the long term strength parameters (c', ϕ').

The analysed soil is overconsolidated owing to its geological history and to past processes (like desiccation and swelling). So the mobilized shear strength should be greater than the long term shear strength developed with great desplacements and dissipated porepressures.

6 Conclusions

At the upper part of slopes in the Carxofes gully, and in other gully tributaries of the Alcoi river, mass movements of mudslide type, which pass into a flow, are common, affecting a layer of weathered material 40–50 cm thick and removing volumes of material up to 100 m^3.

These mudslides are directly related to heavy and copious rainfall, which include important daily storms. These rains are normally preceeded by dry periods.

The zone of initiation of the mudslide studied is concave in plan, where surface and subsurface wash concentrates.

The critical depth of mobilizing shear strength, assuming an infinite slope (GRAHAM 1984), is 2 m and 1.98 m applying Coulomb's model as proposed by JOHNSON (1984).

After the mudslide had occurred concentrated surface and subsurface wash slightly modified the mudslide morphology.

Our proposed explanation of the evolution of the upper slopes of the gully studied is that the concave sections are more active than the convex ones and they evolve through a combination of mudslide and slope wash processes. This differential rate of development may eventually lead to the formation of a tributary gully.

Acknowledgement

We are very grateful to Dr. J. Celma and to the personnel of the "Laboratorio de Mecánica de Suelo del Departamento de Ingienería del Terreno de la Universidad Politécnica de València" for their help in soil analysis and interpretation of these results. We are also very grateful to the reviewers for their valuable critical comments.

References

BRUNSDEN, D. (1984): Mudslides. In: D. Brunsden & D.B. Prior (eds.), Slope Instability. John Wiley & Sons, 363–410.

COLIN ROUSE, W. (1984): Flowslides. In: D. Brunsden & D.B. Prior (eds.), Slope Instability. John Wiley & Sons, 491–522.

CROZIER, M.J. (1973): Techniques for the morphometric analysis of landslide. Z. Geomorph. N.F. **17**, 78–101.

DELAPUE et al. (1964): Les mouvements de masse dans les sols. Rev. Géogr. du Maroc, **6**, Rabat, 29–52.

GRAHAM, J. (1984): Methods of stability analysis. In: D. Brunsden & .B. Prior (eds.), Slope Instability. John Wiley & Sons, 171–215.

JOHNSON, M.A. (1984): Debris Flow. In: D. Brunsden & D.B. Prior (eds.), Slope Instability. John Wiley & Sons, 257–361.

MEZCUA, J. & MARTINEZ SOLARES, J.M. (1983): Sismicidad del área Ibero-Mogrebi. Inst. Geogr. Nacional, Public. **203**, Madrid, 299 pp.

TERZAGHI, K. (1950): Mechanism of landslides. Geol. Soc. Am. Engng. Geol. (Berkey) Vol., 83–123.

YOUNG, A. (1972): Slopes. Oliver & Boyd, Edinburgh. 288 pp.

Address of authors:
Neus La Roca Cervigón and Adolfo Calvo-Cases
Departament de Geografia
Universitat de València
Aptat. 22060
46080 Valéncia (Spain)

SLOPE FORM AND SOIL EROSION ON CALCAREOUS SLOPES (SERRA GROSSA, VALENCIA)

A. Calvo-Cases & N. La Roca Cervigón, Valencia

Summary

This paper studies the slope morphology in a small drainage basin in the South of the province of Valencia. The slope profiles elaborated show different sequences of the most typical forms in this basin as well as in other areas of the País Valenciano, the evolution of which in the Quaternary was characterized by the alternance of phases of downcutting and accumulation at the bottom of the valley. The available data on soil loss rates show the importance of heavy rainfall in areas affected by forest fires, and their evolution after some years.

Resumen

Este artículo estudia la morfología de las laderas de una pequeña cuenca en el Sur de la provincia de Valencia. Los perfiles de ladera elaborados muestran diferentes secuencias de las formas más típicas existentes tanto en esta cuenca como en otras áreas del País Valenciano, cuya evolución durante el Cuaternario ha estado presidida por la sucesión de fases de incisión y relleno del fondo del valle. Los datos disponibles sobre las tasas de pérdida de suelo muestran la importancia de los episodios de lluvias torrenciales en áreas afectadas por incendios forestales, así como su evolución tras algunos años.

1 Objectives and Study Area

The purpose of this paper is to analyze the slope forms on limestones and to reconstruct some aspects of their evolution from present forms. Furthermore, the availability of some data on soil transport by surface wash allows us to know how surface wash acts on the slope forms, particularly during heavy rain periods.

The area of study is situated in the Serra Grossa, an anticlinal structure with Betic orientation (WSW-ENE), which is the western limit of La Vall d'Albaida depression. The relief here is articulated into a series of valleys and divides following the orientation of the mountains and it is rather steep with differences of level of 300 m in only 2 km.

The geological substratum consists of limestones from the Upper Cretaceous, specifically from the Campanian and Maestrichtian series of massive limestones and yellow calcoarenites. In the NW and SE these limestones are covered by more recent materials, from the Tertiary, consisting mainly of Vindobo-

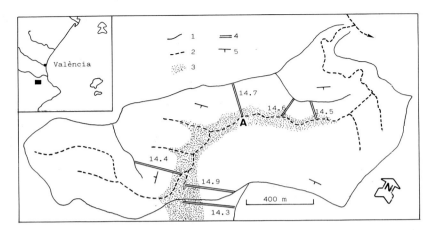

Fig. 1: *Morphological scheme of the Barranc de la Penya de la Mel. 1) Divides; 2) Stream channel; 3) Foot slope deposits; 4) Slope profiles; 5) Dip direction.*

nian marls of the "tap facies" (see IGME 1981). The lithological difference between the mountain (hard rocks) and its sides (soft rocks) will be an important factor in the morphogenesis.

The climate is characterized by temperatures ranging from 10 °C in January to 27°C in July and August. The annual rainfall distribution presents two maxima, one at the end of Autumn and one in Spring. The annual totals are important (mean of 550 mm) but the regime is irregular, with frequent storms; maximum intensities of rain, according to ELIAS & RUIZ (1979), are between 99 and 361 mm/24^h, for periods of recurrence of 2 and 100 years, respectively.

2 Slope Morphology and Evolution

The valley slopes in the middle sector of the Serra Grossa are representative of the calcareous lithology areas of the País Valenciano (CALVO 1986); they contain many signs of the morphoclimatic evolution of the landscape. The drainage basin studied here (fig.1) belongs to a tributary river of the Albaida river, called "El Barranc de la Penya de la Mel". The valley shape is related to the structure which determines a certain asymmetry and irregularities in the drainage basin perimeter. Another significant feature is the great gradient between the highest point of the drainage divide (460 m) and the river mouth at 120 m, a gradient of more than 280 m in 2 km. This high gradient provokes a strong downcutting, particularly in the final 700 m of the main channel. Upwards downcutting becomes progressively negligible and in the upper part there is a small plain filled with deposits. In this sector the valley is wider with smoother slopes and many remains of footslope accumulations.

We have analysed the slope morphology from six longitudinal profiles, whose location can be seen in fig.1. The profiles were measured in the field using the methods described by YOUNG et al. (1974) with a length of measurement of 5 and 10 m. The preferred location

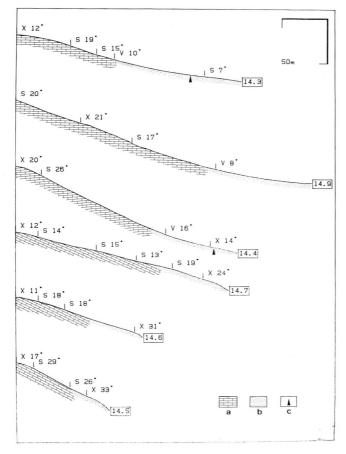

Fig. 2: *Slope profiles.*

a) Bare rock surface; b) Footslope accumulation; c) Sediment traps.

of the profiles on the left bank of the basin results from the need to avoid the distorsions in the form produced by the road which runs along the right bank.

The division of the profiles into morphological units was carried out, in principle, using an objective method, the "Best Units System" from YOUNG (1971). This division was subsequently modified subjectively in order to obtain larger and more general units. This way of delimiting units was chosen in order to avoid the excessive breaking-up of the profile into very small units, the same problem as PARSONS (1977) encountered, and which is due to the high rate of micro-relief in these slopes.

The average slope of the profiles ranges from 12° to 28° and the general profile form is convex-straight-concave or convex, with debris accumulation in the lower half of the profiles; the remainder surface is rocky.

In the same way as it occurs in the average slope, profile form varies according to its drainage basin position. There are two clear morphological types, represented by profiles 14.3 and 14.5, and some intermediate situations, represented by the other four profiles (fig.2).

This morphological series is the result of the progressive channel downcutting along the valley.

Profile 14.3 is representative of the slope form at the head divide; its form consists of five units: one convexity on the crest, two segments at half slope and a great concavity at the foot, followed by a segment which is the result of the change in shape due to the road. Basal removal is impeded, directly related to the concave element. Regolith distribution along this profile shows three clearly different zones:

- The upper part is a bare rock surface with many and well-developed solution microforms (rillenkarren) and some boulders. Hence it is assumed that form is controlled by weathering processes, especially limestone solution and mechanical weathering.

- In the middle part, the upper part of the basal concavity, debris materials appear on the surface including some fragments of a thin laminar crust, exposed in some places. Here the surface undergoes an intense wash erosion: small rills and many stones on soil surface; there is control by detachment.

- The lower part of the basal concavity, third zone, is a place where the dynamic conditions change. The absence of channel downcutting and a low gradient allow the wash material to be held, which gives a net soil gain (control by accumulation). This deposit is made up of fine fraction with little coarse material.

These three zones are well differentiated in the field and affect vegetation showing an increase in density of the more demanding species near the accumulation zone.

Yet with some slight differences, the neighbour profile 14.9 shows the same morphological features. The differences can be seen on the crest, where there are some steps showing the existence of a relict cliff, and therefore there is still some influence of the structure.

Profile 14.4 is the beginning of the progressive morphological change due to channel downcutting. The slope form consists of a small convex crest, a long segment and a concavity, interrupted by the channel, which is responsible for the convex element at the footslope. As a consequence, basal removal is not impeded and, as can be seen in tab.1, there is an important change in the dynamics. The three zones of the previous profiles become two zones, as the accumulation zone does not exist.

The other profiles in this drainage basin (14.7, 14.6, 14.5) show no basal concavity. This is substituted for a convex element with ever steeper gradient. The result is, as seen on the profile further away from the head (14.5), a convex-straight-convex form, with the last unit developed on the remains of the old footslope accumulation (fig.2).

This series of slope profiles shows a set of morphological changes which occurred during the Quaternary, and which gives us important information about the slope evolution in this period. On the different slope sections the lower part undergoes changes most intensively and that debris removal and deposition play an important role. In fact, in all the slope areas of the region studied in a larger work (CALVO 1986) the most characteristic feature of the slope profiles developed on limestones is this convex slope break at the foot. Therefore, the lower part of the slope is of a major importance when reconstructing the evolution sequences, whereas middle and upper parts show very few changes in their form, with the scheme of a convex crest followed by a straight segment.

3 Slope Deposits

Alluvial and colluvial materials at the bottom valley and footslope areas can be grouped into three different sedimentary sets, separated by clear erosion episodes, and therefore they can be considered to be of different ages. We have no direct information about the date of these groups of deposits, but the number of groups and their geometrical and sedimentological characteristics are very similar in most of the areas of a larger region (see CALVO 1986), especially on limestones areas. These three series of footslope deposits are closely related to the river terrace levels of the region, about which chronological information is available from the works of FUMANAL (1986) at the nearby archaeological site of the Cova Negra, and from CARMONA et al. (1986) and PROSZYNSKA-BORDAS (1986).

- The oldest slope sediments (**series I**) consist of a crusted breccia with angular boulders and little matrix. The remains associated with this series are spread out and very small in size. They are always in contact with bedrock at the bottom of the present river bed, in the basin head area, or they are small remains hanging at middle slope in the lower basin area. From the sedimentological point of view sample X-1 in fig.3 shows transport in dense flow with poor classification and a great amount of silt-clay particles in the matrix.

- The **Series II** of footslope deposits is better represented; its distribution coincides with the footslope accumulation area in fig.1. In the stratigraphic profile at point A in fig.1 on the footslope of profile 14.7, the stratigraphic series, 5 m thick, begins with a layer of 1.25 m thick and massive structure with sub-rounded boulders, homometric at about 10 cm and almost without matrix. The boulder imbrication shows at this level a flow in the valley direction; fine particle granulometry (sample X-2) corresponds with typical fluvial environments with good classification. It is a slope deposit redistributed by the channel flow with high transport conditions.

From here to the slope surface the stratigraphic profile consists of an alteration of levels, 25 to 50 cm thick. Some of these (sample X-4) with small subangular boulders, little matrix and massive structure, have similar statistical parameters to sample X-2 and therefore they must have been formed under high flow conditions. Other strata show red silt accumulations comprising some bigger weathered boulders. In this case the sudden decrease in particle average size and worse classification show that wash transport conditions and the supply source could be a previous soil formation. In this context, we have observed in the basin head one remain, 50 cm thick, of red clay with polyedric structure which could be identified

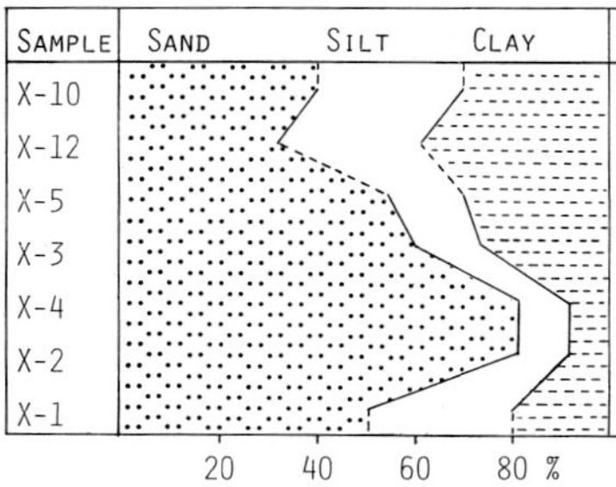

Sample	Sand	Silt	Clay	M	S	SK	K
X-10				5.80	2.85	0.29	0.79
X-12				6.76	3.60	0.11	0.67
X-5				5.10	3.75	0.70	0.76
X-3				4.43	4.20	0.55	0.94
X-4				1.53	2.20	0.46	1.59
X-2				1.76	2.40	0.33	1.29
X-1				4.43	3.20	0.19	0.86

Fig. 3: *Sedimentological analysis of footslope deposits. M = Mean size; S = Standard deviation; SK = Skewness; K = Kurtosis, in phi units.*

as the argilic horizon (Bt) of an old Mediterranean red soil formed at a time of stability after the deposits of the series I.

- The most recent deposits (**Series III**) subsequent to the second aggradational period, consist of a set of small or thin accumulations at those places where material removal is now impeded or slow. These are found in the lower parts of slopes with concave longitudinal profiles (14.3 and 14.9). Materials making up this series are massive accumulations of sandy-clay texture (sample X-5), with scattered boulders. Sedimentary indices are similar to sample X-3, showing variable wash transport energy.

4 Rates of Surface Transport by Wash

The analysis of the form and deposits of these slopes offers some information on the interpretation of the effect of processes on this kind of landscape. It has already been seen how surface wash, and in some cases dense flow, are the most important processes with respect to the aggradation of the oldest deposits. Surface microforms help us to distinguish two sectors in relation to the slope dynamics. The upper part of the slopes is rocky and therefore submitted to control by weathering; solution is perhaps the dominant process, as can be deduced from the width of the diaclases and the presence of some surface rillenkarren. In the central and lower parts of the slopes the conditions are different; rills or microrills and a large number of stones show a net soil loss due to surface wash. Only in slopes with a well developed basal concavity are there some signs of control by accumulation at the foot.

The data on soil transport rates obtained from field measurement show the activity of surface wash and highlight the differences between the dynamics on slopes with basal concavity and on slopes with channel downcutting at the foot, the latter showing different gradient as a consequence of this downcutting.

We have obtained records for five periods on the volume of soil transported by overland flow in two slopes (14.3 and 14.4) at the head divide area of the basin. On both slopes two soil traps were placed in the middle lower slope (fig.2). These traps are large in size (≥ 2 m in mouth

Period	A	B	C	D	E	Total
From	06-May-82	04-Oct-82	30-Dec-83	11-Sep-84	25-Sep-86	
To	03-Oct-82	23-Dec-82	10-Sep-84	02-Nov-85	03-Oct-86	
No. days of rain	9	9	22	34	4	78
Pmm max 24h	69	180	27	39	180	180
Pmm max $\geq$24h	69	256	29	44	389	389
Pmm total	151	318	200	338	389	1396
Slope 14.3						
gr/m	265	14192	110	233	615	15415
gr/m^2 (min)	13	686	5	11	30	745
gr/m^2 (max)	20	1065	8	17	46	1157
mm/1000 years	26	2607	6	8	1129	249
mm/1000 years	40	4048	10	13	1753	387
Slope 14.4						
gr/m	159	126057	1126	385	8025	135752
gr/m^2 (min)	2	1345	12	4	86	1449
gr/m^2 (max)	3	2724	24	8	173	2933
mm/1000 years	4	5337	15	3	3398	505
mm/1000 years	7	10806	30	6	6879	1023

Tab. 1: *Erosion rates from slopes 14.3 and 14.4 and summary of climate conditions.*

width) built up with bricks and concrete and are designed to obtain results in very active or long-term periods of time, and therefore data accuracy and wash water storage have been neglected.

The area suffered a forest fire in the summer of 1981; the traps were installed in November, 1981, and the first data available are from a period that starts in May, 1982.

Soil loss values are shown in tab.1 with the precipitation conditions for each recorded period. Because of the small number of traps the data are difficult to generalize from, but this is the only information available at this moment. In order to transform the soil transport information obtained from the traps into soil loss values as units of weight and surface lowering, the most probable source area for each trap was measured in the field. The field survey was carried out by measuring several cross sections of the slope, using a slope pantometer one meter long, from which a microtopographical map was constructed. As the distance of transport for each measurement period is not known, two values are given: the first corresponds to the maximum rates (max) and it refers to the minimum area estimated on the map; the second value corresponds to the minimum rates (min) and it refers to the maximum area estimated on the map.

The lengths of transport used to calculate the contributing area for each slope were estimated in the field using morphological criteria. These lengths are 17 and 28 m for slope 14.3 and 23 and 36 m for slope 14.4. These values correspond to 10 and 16% of the total slope length from the trap to the divide. Using the data obtained by YAIR & LAVEE (1974) (contributing area between 15–35%) in more arid environments and taking into account the differences of soil properties we think, that the most probable values for heavy storms are the longer lengths, and for low intnsity rains it is better to use the shorter ones.

In 1981, some time after the forest fire, soil properties of the contributing areas of both slopes were studied by PEREZ & SANROQUE (1982) in order to apply the Universal Soil Loss Equation. The K value determined was 0.32 and the results of USLE were 233 and 712 Tm/ha/year for slopes 14.3 and 14.4, respectively. Both slopes had a high percentage of stone cover surface (50 and 60%) and a poor vegetation cover (15 and 10%); the organic content of soil was low (2.4%), compared to the present values of 3.3% on 14.3 and 3.1% on 14.4.

The analysis of the results (tab.1) shows the differences of the soil loss rates for both slopes with different mean gradient and foot removal conditions, as well as the evolution of these rates during the six years following the forest fire.

During period A, the soil transport rate was 1.7 times greater on the slope 14.3, having a less steep gradient (11.9°) but in subsequent periods (B, C, D and E) the inverse relationship between erosion ratios and slope angle no longer exists. It is profile 14.4, with steeper gradients (22.5°) and active basal removal, which shows higher rates. One possible explanation, given by CALVO & FUMANAL (1983) is a greater erodiblity of the soil on slope 14.3, which is in Holocene deposits less compact than the other soils in late Pleistocene deposits.

Another possible explanation for the inverse relationship between gradient and transport rate during period A, can be found in the conclusions stated by YAIR & LAVEE (1981). These authors consider as one of the factors responsible the negative effect of the stone cover on the velocity of very shallow overland flow. This assumption might also explain the similarity of the values on both slopes for record D, a period of low intensity rains.

Together with the different soil compactness and the different stone cover, the contrast between soil erosion rates on both slopes is also related to the infiltration rates; these have been measured with double cylinder infiltrometer. Values for slope 14.4 are four times higher than for slope 14.3.

From the second period onwards, the situation is reversed and soil loss is always greater on the steeper slope. Data for periods B and E are particularly interesting because they correspond to periods of rain with a maximum intensity of 180 mm/24^h (5–10 years of recurrence) and 256 and 389 mm in four consecutive days, respectively. In contrast periods C and D are longer but contain less extreme rainfall events.

The results obtained show some extremely high values, for example for the storm in October, 1982, (period B), which can be compared with some values obtained in badlands (see YOUNG 1974 and SAUNDERS & YOUNG 1983) and are higher than those obtained by YAIR & LAVEE (1981) in more arid environments. However these extreme values only occur during very heavy storms and under conditions of little soil protection.

Low intensity rains produce moderate rates of soil loss and their effect on the environment can even be considered as beneficial, as they allow the rapid growth of some plants.

The differences in transport rates (gr/m) between both slopes increase with time, except for period D, explained above, as we can see in fig.4. The increase ranges from 8.9 times in period B

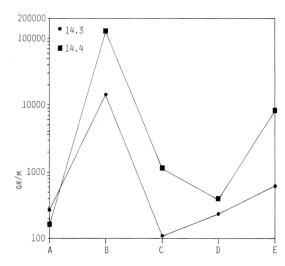

Fig. 4: *Evolution of soil transport rates in the slopes 14.3 and 14.4.*

to 13 times in priod E. This characteristic shows the role of the different velocity of vegetation regeneration due to the effect of gradient on soil erosion.

The data obtained allow a different approach from the point of view of the analysis of average soil loss. This, in mm/1000 years, gives maximum values that range from 387 at 14.3 to 1023 at 14.4. But if we consider a probability of recurrence of maximum rain in 24^h, of seven years for periods B and E, and of one year for the others, these values decrease to 68 and 157 mm/1000 years for each slope. These are similar to the values obtained by FUMANAL & CALVO (1981) from archaeological sediments for the last 5000 years, which range from 44 to 77 mm/1000 years, in a slope with very similar morphology and on the same kind of rock.

Compared to other data available from Mediterranean environments, the value of 0.4 mm/7 months obtained by ORENGO & ROSSI (1973) after forest fire, are higher than the values obtained here in the first record period after the forest fire, although rainfall intensities were also higher in that area. However, the results obtained by LA ROCA (1984) on deforested clay slopes in the interior of Valencia province are lower than our mean values, as they give an average rate of 76 mm/1000 years for a two-year record, including the heavy storms of 1982.

5 Conclusions

Slope evolution, in this environment studied, is very slow; from the middle Pleistocene (probable date of the oldest deposits) until today the changes in slope form have only consisted of an alternation of periods of bottom valley fill, with slope profiles concave at the foot, and periods of channel downcutting at the slope foot, resulting in convex-straight-convex profiles. This characteristic evolution is the same on slopes in other areas having a similar morphoclimatic environment and lithology (CALVO 1986).

Present conditions, heavy rainfall and intense deforestation result in very high soil erosion rates by surface wash, especially on the steeper slopes; these are those slopes having a basal downcutting. In the short-term, these high erosion rates make vegetation regeneration difficult and in the middle- and longterm there is a tendency for the overall removal of the foot slope deposits, as can be seen by the great number of rocky slopes, free faces with 25° gradients, typical of the calcareous areas in these Mediterranean regions.

Acknowledgement

We are very grateful to Dr. A. Pérez Cueva and Dra. M.P. Fumanal García for their help in field work and deposits study. We are also very grateful to M.A. Jaime Pastor for her translation of this paper into English and to two anonymous reviewers for their valuable critical comments.

References

CALVO A. (1986): Geomorfología de laderas en la montaña del País Valenciano. Tesis Docoral. Universidad de Valencia. 344 ff.mec.

CALVO, A. & FUMANAL, M.P. (1983): Repercusiones geomorfológicas de las lluvias torrenciales de octubre de 1982 en la cuenca media del Río Júcar. Cuadernos de Geografía, **32/33**, 101-120.

CARMONA, P., FUMANAL, M.P. & LA ROCA, N. (1986): Paleosuelos pleistocenos en el País Valenciano. In: F. Lopez-Bermudez and J.B. Thornes (eds.), Estudios sobre geomorfología del sur de España, **43-7**, Murcia.

ELIAS, F. & RUIZ, L. (1979): Precipitaciones máximas en España. Ser. Pub. Agrarias. Ministerio de Agricultura, 545 p.

FUMANAL, M.P. (1986): Sedimentología y clima en al País Valenciano. Las cuevas habitadas en el Cuaternario reciente. Servicio de Investigación Prehistórica. Diputación de Valencia, 207 p.

FUMANAL, M.P. & CALVO, A. (1981): Estudio de la tasa de retroceso de una vertiente mediterranea en los últimos 5000 años. Cuadernos de Geografía, **29**, 133–150.

IGME (1981): Mapa geológico de España (escala 1:50.000), Hoja 795 (Játiva). Instituto Geológico y Minero de España. Madrid.

LA-ROCA, N. (1984): La erosión por arroyada en una estación experimental. Cuadernos de Investigación, **X(1-2)**, 85–98.

ORENGO, C. & ROSSI, G. (1973): Sur l'evolution des versants denudes par incendie sous climat méditerranéen. Méditerranée, **1**, 95–105.

PARSONS, A.J. (1977): Curvature and rectilinearity in hillslope profiles. Area, **9**, 245–251.

PEREZ-CUEVA, A. & SANROQUE, P. (1982): Erosión acelerada en vertientes incendiadas. Unpublished paper, 44 ff.mec.

PROSZYNSKA-BORDAS, H. (1986): Thermoluminescence dating of sediments from fossil red soils in the region of Valencia (Spain). In: F. LOPEZ-BERMUDEZ & J.B. THORNES (eds), Estudios sobre geomorfología del sur de España, 113-4, Murcia.

SAUNDERS, I. & YOUNG, A. (1983): Rates of Surface Processes on Slopes, Slope Retreat and Denudation. Earth Surf. Proc. & Landforms, **8(5)**, 473–501.

YAIR, A. & LAVEE, H. (1974): Aereal contribution to runoff on scree slopes in an extreme arid environment. A simulated rainstorm experiment. Z. Geomorph. suppl. Bd. **21**, 106–121.

YAIR, A. & LAVEE, H. (1981): An investigation of source areas of sediment and sediment transport by overland flow along arid hillslopes. Erosion and Sediment Transport Measurement (Proc. Florence Sympos.). IAHS Publ. **133**, 433–446.

YOUNG, A. (1971): Slope profile analysis: the system of best units. Inst. Br. Geogr. Spec. Publ., **3**, 1–13.

YOUNG, A. (1974): The rate of slope retreat. Inst. Br. Geogr., Sp. Publ. **7**, 65–78.

YOUNG, A., BRUNSDEN, D. & THORNES, J.B. (1974): Slope profile survey. B.G.R.G. Tech. Bull. **11**.

Address of authors:
Adolfo Calvo-Cases and Neus La Roca Cervigón
Departament de Geografia
Universitat de Valencia
Aptat. 22060
46080 Valencia (Spain)

THE EFFECT OF CUP SIZE ON SPLASH DETACHMENT AND TRANSPORT MEASUREMENTS PART I: FIELD MEASUREMENTS

J. **Poesen**, Leuven
D. **Torri**, Firenze

Summary

Knowledge of the mass of sediment (MS) detached by drop impact at a soil surface is important for several reasons: i.e. for the estimation of interrill erosion, for the estimation of the rate of depositional crust formation and for the estimation of the protection effectiveness of different canopy covers. MS has often been measured with splash cups used both as receiving collectors or as soil targets. In both cases the estimation of the mass of detached sediment per unit of ground surface (MSA) is strongly influenced by the cup dimension and by the mean splash distance of the sediment. Since no relevant data exist, field measurements, using seven different cup diameters (D) ranging between 0.37 and 24.0 cm, have been carried out to establish the influence of D on MSA.

An exponential relation best describes the field data. The significance of this relation is discussed in relation to the various soil surface properties influencing mean splash distance: i.e. soil shear strength, depth of a surface water film and mean grain or aggregate size. From these considerations it can be concluded that the proposed D-MSA relationship is soil- and surface state-specific. Nevertheless, it allowed us to propose a first correction of published field splash data, used in order to assess the resistance of soils to detachment by raindrop impact.

Résumé

Une connaissance quantifiée de la masse de sol (MS) déplacé en surface par rejaillissement, sous l'impact des gouttes de pluie, présente de multiples intérêts, notamment pour évaleur les pertes en sol sur les interrigoles, la vitesse de formation des croûtes de dépôts et l'effet protecteur de différents couverts végétaux. MS est souvent mesuré au moyen de godets cylindriques placés en surface du sol et utilisés soit comme collecteurs, soit comme sources de sédiments. Dans les deux cas, l'évaluation de la masse de sédiments détachés par unité de surface (MSA) est fortement influencée par les dimensions des godets et par la distance moyenne de rejaillissement des particules et agrégats. Des mesures de terrain ont été effectuées avec des godets de

ISSN 0722-0723
ISBN 3-923381-12-3
©1988 by CATENA VERLAG,
D–3302 Cremlingen-Destedt, W. Germany
3-923381-12-3/88/5011851/US$ 2.00 + 0.25

différents diamètres (D), compris entre 0.37 et 24.0 cm, pour établir l'influence de D sur MSA, des données appropriées n'étant pas disponibles par ailleurs.

Les résultats ont permis d'établir que la relation s'apparente à une fonction de type exponentiel. La signification de la relation est discutée en rapport avec des propriétés de la surface du sol influençant la distance du rejaillissement, telle la résistance au cisaillement, l'épaisseur de la lame d'eau en surface et la taille moyenne des particules ou agrégats. Il en ressort que la relation proposée est dépendante du type de sol et de l'état de la surface: elle permet cependant de proposer une première correction des données déjà publiées concernant la résistance des sols au rejaillissement.

Resumen

El conocimiento de la masa de sedimento (MS) mobilizado por el impacto de las gotas de lluvia en la superficie del suelo es importante por varias razones, como las estimaciones de la erosión entre arroyaderos, de la tasa de formación de costras deposicionales, y de la efectividad de la protección de diferentes coberturas vegetales. La masa de sedimento ha sido a menudo evaluada a partir de cubetas, utilizadas bien como colectores del sedimento salpicado a su alrededor, bien llena de tierra como área muestra de la que es erosionado el sedimento. En ambos casos la estimación de la masa de sedimento arrancado (MSA) por unidad de superficie está fuertemente influenciada por el tamaño de la cubeta y por la distancia media de salpicadura. Puesto que no existen datos apropiados sobre este proceso, hemos llevado a cabo mediciones del mismo en el campo utilizando cubetas de diámetros (D) comprendidos entre 0.37 24.0 cm, a fin de poder establecer al influencia del diàmetro (D) en la masa de sedimento arrancado (MSA).

La mejor descripción de los datos de campo se ha obtenido mediante una relación exponencial. La significación de esta relación está en conexión con varias propiedades del suelo que tienen influencia en la distancia media de salpicadura, como por ejemplo la fuerza de cizallamiento, la profundidad de la lámina de agua superficial, et tamaño medio de los granos y agregados. A partir de estas consideraciones puede concluirse que la relación D/MSA es específica del suelo y del estado de la superficie. No obstante, nos ha permitido proponer una primera correción de los datos publicados sobre salpicadura en el campo y que se utilizan para estimar la resistencia de los suelos a ser disgregados por el impacto de las gotas de lluvia.

1 Introduction

Measurements of the mass of sediment (MS) detached at a soil surface during rainfall have been carried out for several reasons. Since raindrop detachment is the dominant erosion subprocess influencing erosion rates on interrill areas (MEYER et al. 1975), MS needs to be known in order to assess the potential soil mass which can be lost during subsequent transport either by splash droplets (e.g. POESEN 1985) or by surface runoff (FOSTER & MEYER 1975). Also, building a deterministic model of depositional crust formation, requires knowledge of the mass of soil detached by raindrop impact and displaced over short distances (BOIFFIN 1984). MS has also been measured by a number of investigators in their trial to assess the protection effectiveness of different veg-

No	cup diameter (cm)	dimensions of data presented	reference
1	20.0	mass, mass/unit of ground area	FOURNIER (1958, 1967)
2	15.0 10.0 6.0 5.2	mass, mass/unit of ground area	BOLLINNE (1975, 1980)
3	14.0	mass/unit of ground area	SOYER et al. (1982)
4	14.0 10.0 8.0	mass/unit of ground area	MITI et al. (1984)
5	14.0 11.9 11.0 8.9 6.4 5.3	dimensionless	BOIFFIN (1984)
6	10.2	mass, mass/unit of ground area	SREENIVAS et al. (1947)
7	10.0	mass/unit of ground area	SHARMA et al. (1976) SHARMA & PANWAR (1977)
8	7.5	mass	FROEHLICH & SLUPIK (1980)
9	7.0	mass/unit of ground area	ROELS (1984)
10	5.6	mass/unit of ground area	SIDIRAS et al. (1984)
11	3.0	mass/unit of ground area	MORGAN (1983)
12	3.0	mass/unit of rainfall energy	FINNEY (1984)
13	2.1	mass/unit of ground area	POESEN (1985, 1986)
14	-	mass/unit of ground area	GORCHICHKO (1977)

Tab. 1: *Use of cups (funnels) with circular aperture to assess mass of detached sediment under different environmental conditions.*

etation canopies (e.g. BOLLINNE 1975, SHARMA & PANWAR 1977, SOYER et al. 1982, MORGAN 1983, FINNEY 1984).

In order to measure MS, several techniques have been developed in the last decades. One of them consists of embedding a cup or a funnel with a circular aperture in the topsoil in order to collect the detached sediment from the surrounding soil surface. Eventually, a rim of a few mm is left in order to prevent surface runoff from entering the cup. After dividing the mass of collected sediment by the collecting area of the cup, most investigators express their data as a mass of detached sediment per unit of ground area (MSA) (tab.1). In doing so, they assume that the mass of sediment collected in a cup equals the mass of sediment detached at the soil surface with a same area as the collecting cup area.

Among the researchers who used the splashcup technique, BOLLINNE (1975)

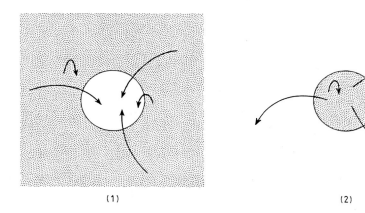

Fig. 1: *Sketch to illustrate the similarity between two splashcup techniques:*

(1) empty splashcup embedded in the topsoil in order to collect detached sediment from the surrounding soil surface, (2) splashcup filled with soil material, detached sediment is collected outside the cup.

was the first to recognize that cupdiameter had a significant negative influence on MSA. This is attributed to the fact that the central part of the splashcup receives less sediment than the parts closer to the cuprim because the mass of deposited splashed sediment decreases exponentially with the distance to the impact point (SAVAT & POESEN 1981, TORRI et al., 1987). Since the area of this central part increases with increasing cupdiameter for a given splash distance, the mass of collected sediment per unit of cupsurface decreases. So, the underestimation of the real MSA (MSAR) increases with increasing cupdiameter. In addition, BOLLINNE (1980) and BOIFFIN (1984) also recognized that underestimation of MSAR, when using a cup of a given size, increased when the splash distance decreased.

From tab.1, it can be seen that investigators used splashcups with diameters ranging between 20.0 and 2.1 cm. From this fact and the preceding discussion it follows that:

1. all the data produced by these investigators are an underestimation of the real mass of detached sediment per unit of soil surface (MSAR);

2. the degree of underestimation is different for each cupsize used and so all the published MSA-data are not mutual comparable.

The problem of underestimation of MSAR is similar for another splashcup technique, i.e. measuring the splash loss from cups filled with soil material (e.g. MAZURAK & MOSHER 1968, MORGAN 1978, SEILER 1980). With this method, part of the detached sediment falls back on the soil surface in the cup and so this part cannot be collected outside the cup. This method is in fact the inverse of that using a collecting cup installed in the topsoil: the source area of one is the collecting area in the other, and vice versa (fig.1).

For both splashcup techniques, theoretical models have been developed enabling one to estimate the underestima-

Splash Detachment and Transport Measurements

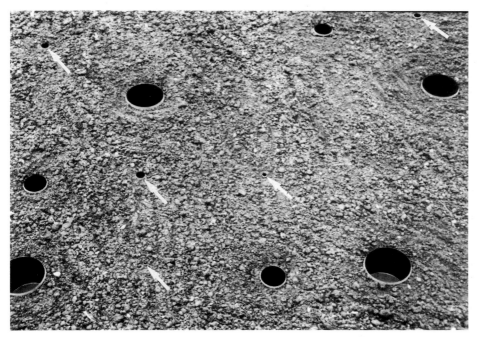

Photo 1: *Random installation of splash cups, having different diameters, in the field. The largest cup diameter equals 19.5 cm.*

tion of MSAR for a given cup size (e.g. FARRELL et al. 1974, REEVE 1982). However, these models have not yet been validated. BOIFFIN (1984) also developed a mathematical model and validated his model with field data. The smallest splashcup he used had a diameter of 5.3 cm. Hence, the assessment of the underestimation of MSAR, when using cups with diameters smaller than 5.0 cm remains questionable. Although other investigators also used cups of different sizes (tab.1), their data do not allow one to calculate the underestimation of MSAR for small cups.

Therefore, field measurements have been undertaken in order to establish a relationship between cup diameter, within the range 0.4 cm–24.0 cm, and mass of collected sediment per unit of cup area. Such a relationship enables one to estimate the correction factor for splash data obtained with circular cups.

2 Materials and Methods

The splashcups used in this study consisted of cylindrical PVC-funnels with a height of 20 cm. Seven different diameters were selected: i.e. 0.37, 0.95, 2.14, 4.34, 10.44, 19.45 and 24.0 cm.

In all, 37 cups were installed vertically in the top layer of a silt loam soil near Tongeren (Eastern Belgium). Maximum surface slope angle equalled 2.6%. Before installation, the soil surface was placed in a conventional seedbed (photo 1), characterized by median dry aggregate size of 0.75 cm as shown in fig.2. Also depicted in the figure is the

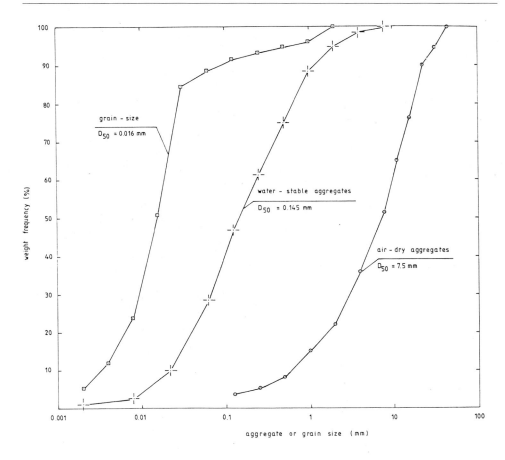

Fig. 2: *Air-dry aggregate size, water-stable aggregate size and grain-size composition of the surface soil.*

water-stable aggregate size (wet sieving method) and the grain-size composition of the surface soil. The marked difference between the two latter distributions is attributed to the binding effect of organic matter, its mean content in the surface soil, determined through loss on ignition, being equal to 5.0%.

Splashcups of different size were installed randomly within a 3 m × 5 m interrill plot (photo 1). Minimum distance between each cup equalled ±50 cm: the fraction of the total mass of detached sediment that splashes more than 50 cm far generally represents less than 5% (SAVAT & POESEN 1981). After installation of the cups, a rim, protruding ±2 mm above the soil surface, was left in order to avoid collection of surface runoff.

During the measuring period, soil surface was kept bare. Leaves and weeds were eliminated from the soil surface without treading on the plot. For this, a scaffold was constructed over the plot. With increasing cumulative rainfall vol-

D (cm)	n	MS (g)	CV	MSA (g cm^{-2})	CV
0.37	8 (+)	0.0080	0.55	0.072	0.57
0.95	5	0.0819	0.35	0.116	0.35
2.14	5	0.2000	1.11	0.055	1.10
4.34	4 (x)	0.9522	0.79	0.064	0.79
10.44	5	2.38	0.28	0.028	0.28
19.45	5	8.08	0.08	0.027	0.07
24.00	2	11.475	0.12	0.0255	0.14

D = splash cup diameter
n = number of splash cups
MS = mean mass of detached sediment collected in one cup
MSA = mean mass of detached sediment per unit of cup surface
CV = coefficient of variation
(+) = data for two cups were omitted
(x) = datum for one cup was omitted

Tab. 2: *Results of field splash measurements.*

ume, cups were pushed deeper into the soil surface when the latter lowered due to soil settling.

At the measuring site, rainfall was recorded daily with two rain gauges, installed at the opposite sides of the interrill plot. Total rainfall amount measured during the period of observation, i.e. from 18.10.84 to 23.12.84, equalled 126 mm. No significant difference in rainfall amount was recorded between the two rain gauges, suggesting that rainfall fell homogeneously over the interrill plot.

Following collection of splashed water and sediment in the field on 23.12.84, partly decomposed organic matter larger than 0.05 cm (i.e. small leaves, worms and insects) were eliminated by sieving. For the cups with a diameter smaller than 4.34 cm, mass of collected splashed sediment was determined with an accuracy of 0.0001 g. For the larger cups, mass of splashed sediment was accurate to within 0.01 g.

3 Results and Discussions

3.1 The Influence of Cup Diameter on MS and MSA

In tab.2 the main results of the field measurements are presented. Beside the mean value of MS and MSA, the coefficient of variation is also calculated. Variation of MS and MSA between the cups reflects not only measuring errors, but also spatial variability of soil surface properties around the splash cups (e.g. aggregate size, microrelief, ...).

As one would expect, MS increases with increasing cup diameter. The relation between D and MS is well described by a power or a linear function (tab.3). A power and an exponential function best represent the negative relation between D and MSA (tab.4). Out of the latter two functions, an exponential one is, from the physical point of view, more sound: e.g. a power function yields an MSA-value of infinity when D equals 0 cm. Hence, an exponential curve is retained to describe our data (fig.3):

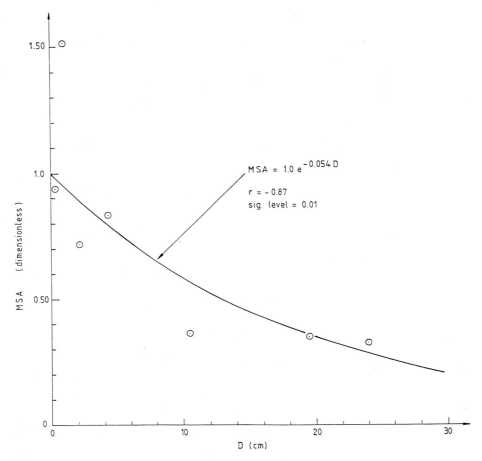

Fig. 3: *Relation between splash cup diameter (D) and mean mass of collected splashed sediment per unit of cup surface (MSA, expressed as a ratio between MSA for a cup with a given diameter and MSA for a cup with D equal to 0 cm).*

$$MSA = 1.0e^{-0.054D} \quad (1)$$

with
- MSA expressed as a ratio between MSA for a cup with a given diameter and MSA for a cup with D = 0 cm, and
- D expressed in cm.

MSA equal to 1.0 represents the real mass of detached sediment per unit of ground area (= MSAR).

FOURNIER (1967), measuring with a cup having a diameter of 20.0 cm, found that on an annual base, 170 ton sediment/ha were detached from a bare sandy ferruginuous tropical soil by raindrop impact in Burkina Faso. If one assumes that the relation given in fig.3 and equation (1) also holds for this type of soil, one can easily deduce that the real mass of detached sediment per soil surface unit equals 500 ton/ha or almost three times the value calculated by FOURNIER.

BOIFFIN (1984) found that the rela-

MS = 0.47	D - 0.86	r = 0.981**
MS = 0.070	$e^{0.24 D}$	r = 0.880**
MS = 2.4	ln D - 0.03	r = 0.822*
MS = 0.060	$D^{1.67}$	r = 0.994**

r = product moment correlation coefficient
** = significant at 0.01 level
* = significant at 0.05 level

Tab. 3: *Statistical relations between cup diameter (D, expressed in cm) and mean mass of detached sediment collected by the cup (MS, expressed in g).*

MSA = −0.003 D + 0.08		r = −0.774*
MSA = 0.077 $e^{-0.054 D}$		r = −0.870**
MSA = −0.017 ln D + 0.079		r = −0.811*
MSA = 0.075 $D^{-0.33}$		r = −0.880

r = product moment correlation coefficient
** significant at 0.01 level
* significant at 0.05 level

Tab. 4: *Statistical relations between cup diameter (D, expressed in cm) and mean mass of detached sediment, per unit of cup surface (MSA, g/cm^2).*

MSA = −0.0001 D^2 + 0.070		r = −0.676
MSA = 0.065 $e^{-0.002 D^2}$		r = −0.775*
MSA = −0.008 ln D^2 + 0.08		r = −0.811*
MSA = 0.075 $(D^2)^{-0.16}$		r = −0.880**

r = product moment correlation coefficient
** significant at 0.01 level
* significant at 0.05 level

Tab. 5: *Statistical relations between cup surface (D^2, in cm) and mean mass of detached sediment per unit of cup surface (MSA, expressed in g/cm^2).*

tion between splash cup surface (D^2) for cup diameters ranging between 5.3 and 14 cm, and MSA, could be well represented by an exponential function. Although this type of equation is physically the most sound, a power or a logarithmic function better fit our data (tab.5).

3.2 Significance of the Field Data

The established relation between D and MSA represents a mean relation for one soil material and a range of surface conditions. The initial surface state was a fine seedbed, characterized by a dry aggregate-size distribution shown in fig.2. During the measuring period, mean aggregate size was reduced and at the end almost 50% of the soil surface was sealed. In addition, water content of the top layer also changed, not only between the rainstorms, but also during rainstorms.

In the introduction it was stated that underestimation of the real mass of detached sediment per unit of cup surface could be explained by the existence of an area, situated around the centre of the receiving area of the cup, which receives less detached sediment than the area closer to the cup rim. Since the size of this central area is inversely proportional to the mean splash distance, it follows that underestimation of MSAR is also inversely proportional to the mean splash distance. The latter is strongly influenced by surface conditions around the cup as will be shown in the following section.

If the resistance of the air to the movement of a splashed particle is neglected, then the splash distance (X) in a horizontal plane can be calculated using the following equation (DE PLOEY & SAVAT 1968):

$$X = \frac{V^2 \sin(2\Theta)}{g} \qquad (2)$$

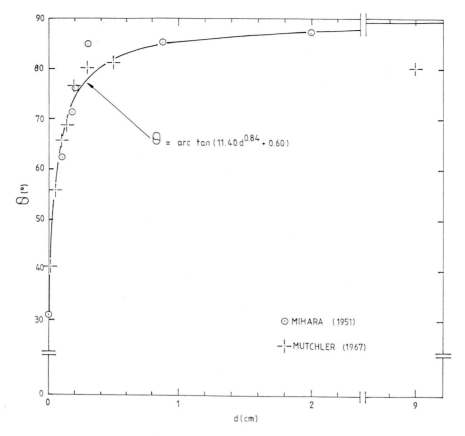

Fig. 4: *Relation bwtween splash angle (Θ) and depth of surface water film (d).*

where
V = the velocity of projection,
Θ = the splash angle (= angle of projection relative to a horizontal plane),
g = the acceleration due to gravity.

AL-DURRAH & BRADFORD (1982) found that, independent of the type of soil material used in their study, the relation between splash angle (expressed in degrees) and shear strength of the top soil (τ, expressed in kPa) could be well represented by the following equation:

$$\Theta = 40.6\tau^{-0.42} \qquad (3)$$

for $1\text{kPa} < \tau < 20\text{kPa}$

Since soil shear strength decreases upon wetting of an initially dry soil top layer during rainfall (POESEN 1981), splash angle and splash distance will probably increase according to equation (3) and (2). Hence, underestimation of MSAR will decrease upon wetting of the soil top layer.

Fig.4 shows the dependence of splash angle on depth of a water film at the soil surface (d, expressed in cm):

$$\Theta = \arctan(11.4d^{0.84} + 0.60) \qquad (4)$$

From this figure, it can be seen that a splash angle of 45°, resulting in a maximum splash distance according to equation (2), corresponds to a water film depth of 0.02 cm. Once this film thickness has been exceeded, splash distance decreases with increasing water depth, a fact already observed in the laboratory (POESEN & SAVAT 1981). Hence, underestimation of MSAR depends on the thickness of the water film at the soil surface.

POESEN & SAVAT (1981) showed that mean splash distance ($\overline{X}$, expressed in m) was also a function of median grain size (D_{50}, expressed in m) of loose sediments:

$$\overline{X} = 0.030(D_{50})^{-0.218} \tag{5}$$

From this observation it can be inferred that aggregate size will also influence splash distance.

From the preceding, one can conclude that some soil surface properties, such as soil shear strength, aggregate size and thickness of a surface water film, have a significant influence on mean splash distance and hence on the underestimation of MSAR. Since it is very likely that these surface properties varied considerably during the measuring period, one can put forward that our field data, presented graphically in fig.3, represent only a mean situation for one soil material and a set of different surface conditions (dry, wet, different surface aggregate size, different microrelief, surface water film with different thickness, ...). Additional data are needed in order to predict underestimation of MSAR for a range of surface materials and surface conditions.

3.3 Use of Established D-MSA Relation to Correct Published Field Splash Data

POESEN (1985) used published field splash cup data to calculate resistance of different soil materials to raindrop detachment (R). These R-values enable one to predict mass of sediment detached on a bare interrill soil surface for a given rainfall event. At that time, however, no data were available to correct the published field data for their underestimation of the real mass of detached sediment per unit of ground area (MSAR).

The relation between cup diameter and MSA, described in this paper, is soil specific. This means that, strictly spoken, it cannot be used to extrapolate or interpolate splash data for soils having a different transportability than the transportability of our field soil. As no other relation exists, our relation is used to correct published R-values. This must be seen as a first correction.

Tab.6 lists the most important data, used in order to calculate Rcorr. MSAR-values were computed with the equation

$$MSAR = MSAe^{0.054D} \tag{6}$$

As stated previously, these corrected field R-values agree well with calculated laboratory R-values (POESEN 1985). Consequently, the Rcorr-values from tab.6 are recommended for use in estimating splash detachment and splash transport on bare interrill soil surfaces.

4 Conclusions

Field splash cup data enabled us to establish an empirical relationship between cup diameter (D) and mass of detached sediment per unit of cup area (MSA) (see fig.3 and equation (1)). With equation

soil texture	D (cm)	KE (J/m^2)	MSA (kg/m2)	MSAR (kg/m^2)	Rcorr (J/kg)	source
clay	10.2	6417	6.42	11.14	576	SREENIVAS et al. (1947) (Texas, U.S.A.)
clay	5.6	1878	3.31	4.48	419	SIDIRAS et al. (1984) (Parana, Brasil)
clay loam	14.0	31654	28.40	60.48	523	SOYER et al. (1982) (Shaba, Zaire)
clay loam	8.0	47410	46.05	70.93	668	MITI et al. (1984) (Shaba, Zaire)
	10.0	47410	31.39	53.86	880	
	14.0	47410	42.89	91.34	519	
silt loam	5.2	7375	41.50	54.95	134	BOLLINNE (1975) (Belgium)
	6.0	3700	2.40	3.32	1115	
loam	7.5	6073	10.74	16.10	377	FROEHLICH & SLUPIK (1980) (Poland)
	7.5	6073	3.99	5.98	1015	
loam	2.1	2940	1.85	2.07	1419	POESEN (1986) (Belgium)
sand	20.0	1886	2.32	6.83	276	FOURNIER (1967) (Burkina Faso)
sand (D_{50} = 0.120 mm)	10.0	17130	103.67	177.90	96	MORGAN (1977) (U.K.)
	10.0	17130	50.00	85.80	200	
sand (D_{50} = 0.192 mm)	2.1	3194	5.00	5.60	570	POESEN (1986) (Belgium)
Stony silty clay (stone cover = 60%)	7.0	6796	0.89	1.30	5232	ROELS (pers. comm.) (France)

D = splash cup diameter
KE = kinetic rainfall energy
MSA = mean mass of detached sediment per unit of ground area
MSAR = real mass of detached sediment per unit of ground area, calculated with eq. (6)
Rcorr = corrected mean resistance of the soil surface to detachment by raindrop impact (= KE/MSAR)
D_{50} = median grain size of the soil material

Tab. 6: *Calculation of resistance to raindrop detachment for bare interrill soil surfaces, using corrected splash cup data.*

(6), the real mass of detached sediment per unit of cup area and also per unit of ground area (MSAR) can be computed using D and MSA as an input.

From the established relation, one is inclined to conclude that in order to measure splash detachment per unit of ground area, the best splash cup size is a very small one. From our experience, however, we recommend to measure splash detachment in the field with a cup having a diameter large enough, i.e. larger than 10 cm, in order to reduce rim effects. Another advantage of a large cup diameter is that collected splash data reflect better the mean detachability of the soil surface than data obtained with small cups. The obtained MSA-data must then be transformed into MSAR-data using an appropriate D-MSA relation.

Since a D-MSA relation is also a function of splash distance, our established relation is soil-and surface state-specific. Hence, this relation can, strictly spoken, only be used for soils with the same transportability as the transportability of the soil for which field data were collected. Nevertheless, the relation found gives a first indication of the underestimation of MSAR when measuring splash detachment from bare soils with a given cup diameter.

More information is needed on the D-MSA relations for soils having a different transportability. Since field and laboratory measurements are rather time-consuming, the construction of a mathematical model, enabling one to predict this relation as a function of different splash distances, would be a more efficient procedure. The field data reported in this paper can then be used to validate such a model.

Acknowledgements

We are grateful to Mrs. C. Poesen for collecting the pluviometric data in the field, Mr. L. Cleeren, Mr. R. Geeraerts and Mr. J. Meersmans are thanked for their technical assistance. We also wish to thank Mrs. A. Van Elsen for typing the manuscript.

References

AL-DURRAH, M. & BRADFORD, J. (1982): The mechanism of raindrop splash on soil surfaces. Soil Science Society of America Journal, **46**, 1086–1090.

BOIFFIN, J. (1984): La dégradation structurale des couches superficielles du sol sous l'action des pluies. Unpubl. Ph.D. thesis, I.N.R.A. Paris-Grignon, 320 p.

BOLLINNE, A. (1975): La mesure de l'intensité du splash sur sol limoneux. Mise au point d'une technique de terrain et premiers résultats. Pedologie, **25**, 199–210.

BOLLINNE, A. (1980): Splash measurements in the field. In: DE BOODT, M. & GABRIELS, D., Assessment of Erosion, Wiley, Chichester, 441–453.

DE PLOEY, J. & SAVAT, J. (1968): Contribution à l'étude de l'érosion par le splash. Zeitschrift für Geomorphologie, **2**, 174–193.

FARRELL, D., MOLDENHAUER, W. & LARSON, W. (1974): Splash correction factors for soil erosion studies. Soil Science Society of America Proceedings, **38**, 510–514.

FINNEY, H. (1984): The effect of crop covers on rainfall characteristics and splash detachment. J. Agric. Engng. Res., **29**, 337–343.

FOSTER, G. & MEYER, L. (1975): Mathematical simulation of upland erosion by fundamental erosion mechanics. USDA-ARS-S-40, 190–207.

FOURNIER, F. (1958): L'Erosion du sol dans les territoires Français d'outre-mer. Publ. IAHS, **43**, 72–75.

FOURNIER, F. (1967): La recherche en érosion et conservation des sols dans le continent Africain. African Soils, **12**, 5–51.

FROEHLICH, W. & SLUPIK, J. (1980): Importance of splash in erosion process within a small flysch catchment basin. Studia Geomorphologica Carpatho-Balcanica, **14**, 77–112.

GORCHICHKO, G. (1977): Device for determining the amount of soil splashed by raindrop. Soviet Soil Science, **8**, 610–613.

MAZURAK, A. & MOSHER, P. (1968): Detachment of soil particles in simulated rainfall. Soil Science Society of America Proceedings, **32**, 716–719.

MEYER, L., FOSTER, G. & RÖMKENS, M. (1975): Source of soil eroded by water from upland slopes. USDA-ARS-S-40, 177–189.

MIHARA, Y. (1951): Raindrop and soil erosion. Bulletin of the National Institute of Agricultural Sciences, A **1**, 1–51.

MITI, T., SOYER, J. & ALONI, K. (1984): Splash en milieux subnaturels de région tropicale (Shaba, Zaïre). Zeitschrift für Geomorphologie Suppl.-Bd. **49**, 75–86.

MORGAN, R. (1977): Soil erosion in the United Kingdom: field studies in the Silsoe Area, 1973–75. Nat. Coll. Agr. Engng. Silsoe Occ. Paper **4**, 41 p.

MORGAN, R. (1978): Field studies of rainsplash erosion. Earth Surface Processes, **3**, 295–299.

MORGAN, R. (1983): Effect of corn and soybean canopy on splash detachment. Paper as Appendix I to Report on NATO research grant 051/82, 21 p.

MUTCHLER, C. (1967): Parameters for describing raindrop splash. Journal of Soil and Water Conservation, **22**, 91–94.

POESEN, J. (1981): Rainwash experiments on the erodibility of loose sediments. Earth Surface Processes and Landforms, **6**, 285–307.

POESEN, J. (1985): An improved splash transport model. Zeitschrift für Geomorphologie, **29**, 193–211.

POESEN, J. (1986): Field measurements of splash erosion to validate a splash transport model. Zeitschrift für Geomorphologie, Suppl. Bd. **58**, 81–91.

POESEN, J. & SAVAT, J. (1981): Detachment and transportation of loose sediments by raindrop splash. Part II Detachability and transportability measurements. CATENA **8**, 19–41.

REEVE, I. (1982): A splash transport model and its application to geomorphic measurement. Zeitschrift für Geomorphologie, **26**, 55–71.

ROELS, J. (1984): Modelling soil losses from the Ardèche rangelands. CATENA, **11**, 377–389.

SAVAT, J. & POESEN, J. (1981): Detachment and transportation of loose sediments by raindrop splash. Part I The calculation of absolute data on detachability and transportability. CATENA, **8**, 1–17.

SEILER, W. (1980): Messeinrichtungen zur quantitativen Bestimmung des Geoökofaktors Bodenerosion in der topologischen Dimension auf Ackerflächen im Schweizer Jura. CATENA, **7**, 233–250.

SHARMA, S., GUPTA, R. & PANWAR, K. (1976): Concept of crop protection factor evaluation in soil erosion. Indian J. Agr. Res., **10**, 145–152.

SHARMA, S. & PANWAR, K. (1977): Effectiveness of crop-cover for reducing splash erosion. Soil Conservation Digest, **5**, 1–7.

SIDIRAS, N., ROTH, C. & DE FARIAS, G. (1984): Effect of rainfall intensity on splash erosion and runoff under three tillage systems. Revista Brasileira de Ciência do Solo, **8**, 251–254.

SOYER, J., MITI, T. & ALONI, K. (1982): Effets comparés de l'érosion pluviale en milieu péri-urbain de région tropicale. Revie de Géomorphologie Dynamique, **31**, 71–80.

SREENIVAS, L., JOHNSTON, J. & HILL, H. (1947): Some relationsips of vegetation and soil detachment in the erosion process. Soil Science Society Proceedings, **12**, 471–474.

TORRI, D., SFALANGA, M. & DEL SETTE, M. (1987): Splash detachment: runoff depth and soil cohesion. CATENA, **14**, 149–155.

Addresses of authors:
J. Poesen
Research Associate
National Fund for Scientific Research
Catholic University of Leuven
Laboratory of Experimental Geomorphology
Redingenstraat 16bis
B 3000 Leuven, Belgium
D. Torri
C.N.R. Centro di Studio per la Genesi, Classificazione, e Cartografia del Suolo
Piazzale delle Cascine 15
Firenze, Italy

THE EFFECT OF CUP SIZE ON SPLASH DETACHMENT AND TRANSPORT MEASUREMENTS PART II: THEORETICAL APPROACH

D. **Torri**, Firenze
J. **Poesen**, Leuven

Summary

The interpretation of field and laboratory data on the mass of sediment detached by raindrop impact per unit of ground surface (MSAR) measured with the splash cup techniques, is often problematic (Part I). Therefore, this paper approaches the problem from a theoretical point of view.

A mathematical model, taking into account the exponential decrease of the mass of deposited splashed sediment with distance from the impact area, has been developed. The model allows the calculation of the mean jump length and of the mass of sediment collected in splash cups of different sizes.

The results of the model have been succesfully compared with measured field data. As the equations can be solved only by numerical approximation, a nomograph is presented allowing the estimation of the mean splash distance and of MSAR. Also recommandations for the use of splash cups in the field are given.

ISSN 0722-0723
ISBN 3-923381-12-3
©1988 by CATENA VERLAG,
D-3302 Cremlingen-Destedt, W. Germany
3-923381-12-3/88/5011851/US$ 2.00 + 0.25

Résumé

La mesure, au moyen de godets cylindriques, de la masse de sédiments détachés par unité de surface (MSAR) sous l'action du rejaillissement pose un problème d'interprétation (cf. Part I): cet article a pour objectif de proposer une approche théorique de cette question.

Le modèle mathématique développé tient compte de la diminution exponentielle du dépôt des sédiments déplacés par rejaillissement avec la distance au point d'impact des gouttes de pluie. Il permet de calculer la distance moyenne parcourue par les particules ou agrégats détachés et la masse collectée dans des godets de différentes tailles.

Les résultats obtenus avec le modèle se sont avérés semblables aux mesures effectuées sur le terrain. Des approximations numériques suffisant à résoudre les équations, il a été possible d'établir un nomogramme permettant d'estimer la distance moyenne du rejaillissement et la masse détachée par unité de surface (MSAR). Certaines recommandations sont également fournies pour l'utilisation des godets de mesure de rejaillissement.

Resumen

La interpretación de los datos obtenidos en el campo y en el laboratorio sobre la masa de sedimento arrancado por el impacto de las gotas de lluvia por unidad de superficie (MSAR), medido utilizando la técnica de las tazas de salpicadura, resulta a menudo problemática (parte I). Por tanto en este artículo se aborda el problema desde un punto de vista teórico.

Se ha desarrollado un modelo matemático tomando en consideración la disminución exponencial de la masa de sedimento arrancado en relación a la distancia desde el área del impacto de la gota de lluvia. El modelo permite calcular la media de la longitud del salto del sedimento y de la masa de sedimento recogido en tazas de diferente tamaño.

Los resultados obtenidos con el modelo han sido comparados con éxito con les mediciones de campo. Puesto que las ecuaciones pueden resolverse solamente mediante aproximación numérica, presentamos un nomógrafo que permite estimar la distancia media del impacto y de la MSAR. También se dan recomendaciones sobre la utilización de tazas en el campo.

1 Introduction

Splash cups, used for measuring splash detachment and transport, have been often misused (POESEN & TORRI, Part I.)

POESEN & SAVAT (1981) found that the mass of soil splashed beyond a certain distance decreases exponentially with increasing distance from the drop impact area. This means that the behaviour of the splashed mass with distance can be described by a sediment decay function. RIEZEBOS & EPEMA (1985) and TORRI et al. (1987) confirmed their findings.

The main consequence of the sediment decay with distance is that any measure of splash detachment is influenced by the mean jump length ($\bar{x}$) of the splashed sediment. Particularly $\bar{x}$ and the splash cup size interfere giving rise to different estimates of the amount of splashed sediment per unit soil surface. FARREL et al. (1974) and REEVE (1982) already worried about those interferences. They proposed some possible equations to solve the interference but they did not base their solutions on any observed sediment decay function. Consequently in this paper an analytical model, based on the observed sediment decay function, is developed and subsequently validated with field data.

2 Splash Cup Model

The equations that will be developed in the following paragraphs refer to the case of a receiving splash cup. The final results also hold for the case of an ejecting splash cup (e.g. MORGAN 1981) for the reasons already discussed by POESEN & TORRI (Part I).

From the theory developed by SAVAT & POESEN (1981), the splashed soil mass jumping beyond a distance ρ from the drop impact area between the angles φ and $\varphi + d\varphi$ (fig.1) is:

$$dA(\rho, \varphi) = \frac{a}{2\pi} e^{-b\rho} d\varphi \qquad (1)$$

where
$A(\rho, \varphi)$ = total mass of soil splashed beyond ρ(g);
a = total splashed soil mass (g);
b = inverse of the mean jump length $\bar{x}(cm^{-1})$.

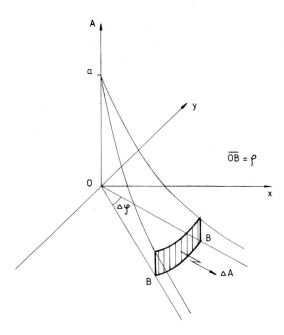

Fig. 1: *Exponential decrease of the mass of splashed sediment (A) with distance (x,y).*

In order to calculate the amount of sediment collected by a cup of radius r (fig.2) only the mass of sediment falling inside the cup — whose centre is at a distance σ from the drop impact area (represented by the point B of fig.2) — must be taken into account. This means that φ ranges between φ_{min} and φ_{max} while ρ ranges between $\rho_C(\sigma,\varphi)$ and $\rho_D(\sigma,\varphi)$. The amount contributed by the source B is then given by the following equation:

$$A_B = \frac{a}{2\pi} \int_{-\varphi_{min}}^{\varphi_{max}} (e^{-b\rho_C} - e^{-b\rho_D}) d\varphi \qquad (2)$$

Let us now explicit the dependence of ρ_C and ρ_D and σ. In order to do that, one needs to find the coordinates of D and C. This means that the following equations have to be solved:

$$\left. \begin{array}{rcl} y &=& m(\sigma - x) \\ x^2 + y^2 &=& r^2 \end{array} \right\} \qquad (3)$$

where the first equation describes the family of straight lines passing through B (m being the slope of the straight line) while the second one describes the equation of the circumference of the splash cup.

The solutions are:

$$\left. \begin{array}{rcl} x_{C,D} &=& \frac{\sigma m^2 \pm \sqrt{(1+m^2)r^2 - m^2\sigma^2}}{1+m^2} \\ y_{C,D} &=& m(\sigma - x_{C,D}) \end{array} \right\} \qquad (4)$$

Hence:

$$\left. \begin{array}{rcl} \rho_C &=& \sqrt{(\sigma - x_C)^2 + y_C^2} \\ \rho_D &=& \sqrt{(\sigma - x_D)^2 + y_D^2} \end{array} \right\} \qquad (5)$$

Now the limits of integrations φ_{min} and φ_{max} are needed. These values correspond to the condition under which the straight line passing through B is tangent to the splash cup. It can be found that:

$$\varphi_{max} = \varphi_{min} = \arctan(r/\sqrt{\sigma^2 - r^2}) \qquad (6)$$

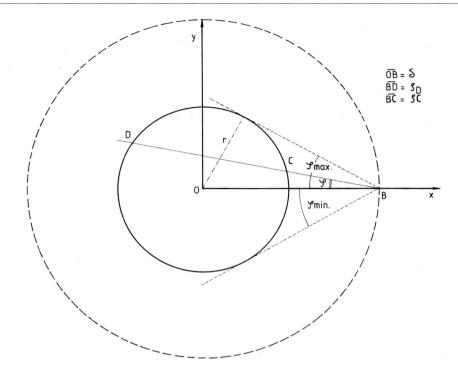

Fig. 2: *Splashcup of radius r and geometrical position of sources placed at a distance σ from the splashcup centre 0.*

Using equations (5) and (6), eqn. (2) can be solved. Obviously eqn. (2) takes into account only the amount contributed by a single point B. All the points placed at distance σ from the centre 0 contribute the same amount. Consequently the total contribution from all those sources at σ from 0 is proportional to the circumference $2\pi\sigma$:

$$A(\sigma) = Ka\sigma \int_{-\varphi_{min}}^{\varphi_{max}} (e^{-b\rho_C} - e^{-b\rho_D})d\varphi \quad (7)$$

where K = constant of proportionally (inverse of a length).

The total mass of collected sediment (MS) contributed by the area surrounding the splash cup can be calculated integrating $A(\sigma)$ over all the possible values of σ, namely r and ∞:

$$MS = K \int_{r}^{\infty} A(\sigma)d\sigma \quad (8)$$

The constant K in eqn. (8) has the same meaning and value as in eqn. (7).

In reality, once σ has reached a value close to 4–5 times the mean jump length, its maximum value being of the order of 20–25 cm (POESEN & SAVAT 1981), the contribution is almost negligible. Hence eqn. (8) can be written as follows:

$$MS = aK^2 \int_{r}^{r+100} \int_{-\varphi_{min}}^{\varphi_{max}}$$

$$(e^{-b\rho_C} - e^{-b\rho_D})\sigma d\varphi d\sigma \quad (9)$$

If we state that:

$$I(b,r) = \int_r^{r+100} \int_{-\varphi_{min}}^{\varphi_{max}} (e^{-b\rho_C} - e^{-b\rho_D})\sigma d\varphi \sigma \quad (10)$$

eqn. (9) becomes:

$$MS = aK^2 I(b,r) \quad (11)$$

K has been introduced as a constant of proportionality between the number of sources and the segment length to which these sources belong: e.g.

$$n(\sigma) = K(2\pi\sigma) \quad (12)$$

where n = number of sources situated on the corcumference of radius σ.

This corresponds to the implicit assumption that the surface density (N) of sources is constant:

$$N = K^2 \quad (13)$$

In reality, the number of sources depends on the total number of drops. Hence, N can be expressed as a function of total rainfall mass per unit of surface (Q):

$$N = \frac{Q}{m_d} \quad (14)$$

where m_d = mass of a representative drop.

Introducing eqn. (14) into eqn. (13) it follows:

$$K^2 = \frac{Q}{m_d} \quad (15)$$

Now eqn. (11) can be rewritten as:

$$\frac{MS}{Q} = a' I(b,r) \quad (16)$$

where

$$a' = a/m_d$$

Obviously the integral $I(b,r)$ (eqn. (10)) can be solved only by numerical approximation.

3 Comparison of the Theory with the Field Data

For a given soil two parameters have to be evaluated, namely a and b.

Let us suppose that we have a set of n splash cups of different size. Using eqn. (10) one can calculate $I(b,r)$ for the n cups and for z values of b.

For each b-value the following ratio can be defined:

$$MSAR_i(b) = MS(b,r_i)/I(b,r_i) \quad (17)$$

where $MSAR_i = a_i(b)K^2$ (eqn. 11).

For each set of $MSAR$-values, mean $(\overline{MSAR}(b))$ and standard deviation (S_b) can be calculated. Comparing the z values of S_b, the $MSAR(b)$ and b corresponding to the smallest S_b-value can be selected as best estimates of the true $MSAR$- and b-value.

In tab.1 the data collected by BOLLINNE (1975) and by POESEN & TORRI (Part I) are listed together with the main outputs of the model. In fig.3 measured versus estimated MS-data are reported. They show a small scatter around the line of perfect agreement, confirming the model.

4 How to Use the Model?

In fig.4 some values of $I(b,r)$ are given as a function of $1/b$ for different values

r (cm)	MS (g)	I(b,r)* (cm²)	$\overline{MSAR}(b)$** ± S_b (g/cm²)	$\overline{x}$ ±Δx (cm)	Source
3.0	6.78	10.78			
5.0	12.03	19.57	0.624	2.0	BOLLINNE
7.5	18.97	30.20	0.0081	0.5	(1975)
0.19	0.01	0.14			
0.48	0.08	0.67			
1.07	0.20	3.13	0.079	6.0	POESEN
2.17	0.95	11.29	0.0212	0.5	&
5.22	2.38	45.96			TORRI
9.73	8.08	105.38			(Part I)
12.00	11.47	135.29			

* I(b,r) is calculated with eqn. (10)
** $\overline{MSAR}(b)$ is calculated with eqn. (17)

Tab. 1: *Splash cup data and parameter values deduced with the model.*

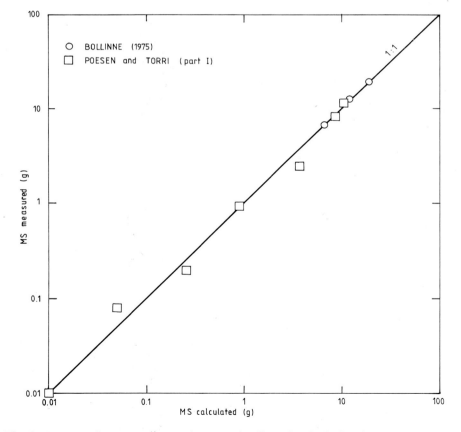

Fig. 3: *Measured versus estimated mass of collected splashed sediment in cups (MS).*

Splash Detachment and Transport Measurements

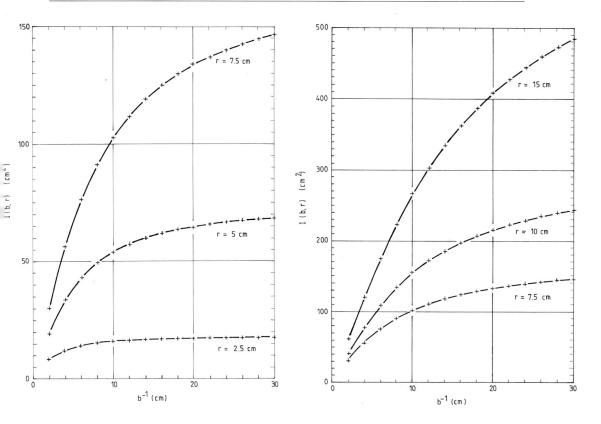

Fig. 4: *Relation between mean jump length (b^{-1}) and the integral (I, equation 10) for different splashcup radii (r).*

r (cm)	MS (g)	1/b (cm)				
		16	8	2	6	4
				I (b,r) (cm^2)		
2.5	2.7	17.3	15.0	8.5	14.0	12.5
5.0	8.2	62.0	50.0	19.5	45.0	34.0
10.0	16.5	196.0	130.0	41.0	110.0	77.0
$\overline{MSAR}(b)$ (eqn. (17))		0.124	0.157	0.380	0.175	0.224
±S_b		0.037	0.027	0.055	0.022	0.015
(g cm^{-2})						

Tab. 2: *Example on the use of the nomograph (fig.4).*

of r. In order to illustrate the use of these curves let us suppose that splash has been measured with three cups, resulting in the data given in tab.2, columns 1 and 2. Nothing is known about the mean jump length. Hence we can select three values of $1/b$ and read the corresponding I-values from fig.4. The first three values of $1/b$ and of I are given in tab.2, columns 3, 4 and 5. The smallest value of standard deviation correspond to $1/b = 8$ cm. Now two other values of $1/b$ can be selected. The values are shown on tab.2, columns 6 and 7. The standard deviation at $1/b = 4$ cm is the smallest. Consequently, the corresponding MSAR-value is selected. It is useless going deeper into detail, trying e.g. with $1/b = 3$ or 5 cm as the errors inherent to the measurements and to the readings of the curves do not allow a better estimate of $1/b$ and of MSAR.

Alternatively, the program given in Appendix II can be used. In order to run the program, number of splash cups, r, MS and estimate of a minimum mean jump length ($\bar{x}$) have to be given. The output will be $\bar{x}, MSAR, S_b$, MS-measured and MS-estimated.

5 Remarks on the Use of the Proposed Technique

The proposed model is based on the assumption of a homogeneous distribution of splashed particles in every direction. Consequently it cannot be used when this assumption is far from being fulfilled: e.g., steep sloping soil surface, herbaceous cover, etc.

For soil material characterized by a mean jump length of less than 2 cm the model is not reliable anymore because the mean jump length is of the same order of magnitude of the diameter of the impact crater.

Special care has to be given to the protrusion of the splash cup rim: periodical inspections should be made in order to keep the rim protruding not more than 2–3 mm above the soil surface, as described by POESEN & TORRI (Part I).

Moreover, the selected splash cup sizes should cover a sufficiently wide range. For instance, a combination such as r=2.5, 5, 10 cm is a good one. Another combination such as r=4, 5, 7 cm (e.g. MITI et al. 1984) is not as good as the previous one because the experimental errors can be too large in comparison with the variation of observed MS-values.

6 Conclusions

After comparing the results of the proposed splash cup model with the measured field data, it can be concluded that the model gives a reliable estimate of the mass of detached sediment per unit of ground area and mean jump length. Consequently, a method is available for a correct interpretation of field splash cup data in the future.

It has to be stressed that the integrals (I) hold for both ejecting and receiving splash cups.

Acknowledgement

The authors are grateful to Dr. J. Savat (†) who recognized the importance of the interference between splash transport distance and cup size in splash measurements. He gave the main basis on which the present model has been developed.

We also wish to thank Ms. A. Van Elsen for typing the manuscript and Mr.

R. Geeraerts for drawing the illustrations.

References

BOLLINNE, A. (1975): La mesure de l'intensité du splash sur sol limoneux. Mise au point d'une technique de terrain et premiers resultats. Pedologie, **25**, 199–210.

FARREL, D., MOLDENHAUER, W. & LARSON, W. (1974): Splash correction factors for soil erosion studies. Soil Science Society of America Proceedings, **38**, 510–514.

MITI, T., SOYER, J. & ALONI, K. (1984): Splash en milieux subnaturels de région tropicale (Shaba, Zaïre). Zeitschrift für Geomorphologie Suppl. Bd. **49**, 75–86.

MORGAN, R.P.C. (1981): Field measurement of splash erosion. Symposium on Erosion and Sediment Transport Measurement. IAHS-AISH Publ. **133**, 373–382.

POESEN, J. & SAVAT, J. (1981): Detachment and transportation of loose sediments by raindrop splash. Part II. Detachability and transportability measurements. CATENA, **8**, 19–41.

POESEN, J. & TORRI, D. (1988): The effect of cup size in splash detachment and transport measurements. Part I: Field measurements. CATENA SUPPLEMENT **12**, 113–126.

REEVE, I. (1982): A splash transport model and its application to geomorphic measurement. Zeitschrift für Geomorphologie, **26**, 55–71.

RIEZEBOS, H.T. & EPEMA, G.F. (1985): Drop shape and erosivity. Part II: Splash detachment, transport and erosivity indices. Earth Surface Processes and Landforms, **10**, 69–74.

SAVAT, J. & POESEN, J. (1981): Detachment and transportation of loose sediments by raindrop splash. Part I: the calculation of absolute data on detachability and transportability. CATENA, **8**, 1–17.

TORRI, D., SFALANGA, M. & DEL SETTE, M. (1987): Splash detachment: runoff depth and soil cohesion. CATENA **14**, 149–155.

Appendix I. List of Symbols

A	mass of splashed soil (g)
a	total splashed soil mass (g)
b	inverse of the mean jump length $(= 1/\bar{x})$ (cm^{-1})
I	integral (eqn. 10) (cm^2); equals the weighted soil surface contributing sediment to the splash cup.
K	constant of proportionality —
	linear density of sources (cm^{-1})
N	surface density of sources (cm^{-2})
n	number of sources
m	slope of a straight line
m_d	mass of a representative drop (g)
MS	mass of detached sediment collected in a splash cup (g)
$MSAR$	real mass of the splashed sediment per unit of surface (g·cm^{-2})
Q	rain mass per unit of surface (g·cm^{-2})
r	cup radius (cm)
S_b	standard deviation of MSAR (g·cm^{-2})
$\bar{x}$	mean jump length (cm)
Δx	interval of mean jump length between two consecutive calculations (see program in appendix I). In fact, Δx represents the approximation by which the mean jump length is known (cm)
x, y	cartesian coordinates
σ	distance between the centre of a splash cup and a drop impact area (cm)
$\varphi, \varphi_{min}, \varphi_{max}$	angles (fig.2)
ρ, ρ_C, ρ_D	distances from an impact point (cm)

Appendix II: Computer Program

```
        DIMENSION X(100),DINT(100),SINT(100),RA(100),AR(100)
        WRITE(7,97)
97      FORMAT(' N.SPLASH CUPS:')
        READ(5,*)N
        WRITE(7,110)
110     FORMAT(' CUP RAYS (CM) - - COLLECTED AMOUNT(G)')
        DO1 K=1,N
        READ(5,*) RA(K),AR(K)
1       CONTINUE
        ZX=1000000000.
        ZM=0.
        WRITE(7,200)
200     FORMAT(' MINIMUM JUMP (CM)')
        READ(5,*)XMIN
        DO5 J=1,28
5       X(J)=XMIN+(J-1)+.5
        DO30 K=1,15
        B=1./X(K)
        OO40L=1,N
        R=RA(L)
        DMIN=R+0.005
        DOD=0.
        DDEL=.0625
        DMAX=5.*X(K)+K
        J1=INI((DMAX/DMIN)/DDEL)
        DO20 J=1,J1
        D=DMIN+(J-1)*DDEL
        TMAX=R/SORT(D*D-R*R)
        TMAX=ATAN(TMAX)
        TDEL=0.026
        JI=INT(TMAX/TDEL)
        BOD=0.
        DO10 I=1,JT
        F1=(I-1)*TDEL
        T1=SIN(F1)/COS(F1)
        T=T1*T1
        DFI=SORT((1.+T)*R*R-T*D*D)
        XC=(D*T-DEL)/(1.+T)
        XC=T1*(D-XC)
        ROC=SORT((D-XC)**2.+YC*YC)
        XD=(D*T+DEL)/(1+T)
        YD=T1*(D-XD)
        RODSORT((D-XD)**2.+YD*YD)
        WRITE(7,*)I,J,L,K
        BOD=(EXP(-B*ROD)-EXP(-B*ROD))+BOD
10      CONTINUE
        BOD=BOD*TDEL2.
        DOD=D*DDEL*BOD+DOD
20      CONTINUE
        SINT(L)=DOD
40      CONTINUE
        SX=0.
        XA=0.
        DO21I=1,N
        XM=XM+AR(I)/SINT(I)
        SX=SX+AR(I)/SINT(I)
```

```
21      CONTINUE
        XM=XM/N
        SX=SX-XM*XM*N
        SX=SORT(SX/FLOAT(N-1))
        IF(SX-ZX)60,60,50
60      ZX=SX
        ZM=XM
        DO23I=1,N
        DINT(I)=SINT(I)
23      CONTINUE
        WRITE(6,400)
400     FORMAT('- - - - - - - - - - - - - -')
        GOTO30
50      WRITE(6,211)X(K-1),ZM,ZX
211     FORMAT(' JUMP (CM)=',F5.2,10x,'A (G)=',F10.3,7X,'SA=',F10.4)
        WRITE(6,206)
        DO301I=1,N
        B=ZM*DINT(I)
        WRITE(6,203)AR(I),B
301     CONTINUE
203     FORMAT(2X,F20.2,F20.2)
206     FORMAT(10X,'MEASURED',7X,'- -',7X,'ESTIMATED')
        K=16
30      CONTINUE
        STOP
        END
```

Addresses of authors:
D. Torri
C.N.R. Centro di Studio per la Genesi,
Classificazione e Cartografia del Suolo
Piazzale delle Cascine 15
Firenze, Italy
J. Poesen
Research Associate
National Fund for Scientific Research
Catholic University of Leuven
Laboratory of Experimental Geomorphology
Redingenstraat 16bis
B-3000 Leuven, Belgium

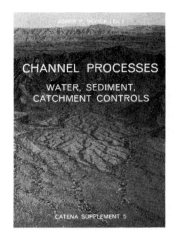

Asher P. Schick (Ed.):

CHANNEL PROCESSES
WATER, SEDIMENT, CATCHMENT CONTROLS

CATENA SUPPLEMENT 5, 1984

Price DM 110,—

ISSN 0722–0723 / ISBN 3-923381-04-2

PREFACE

Two decades ago, the publication of 'Fluvial Processes in Geomorphology' brought to maturity a new field in the earth sciences. This field – deeply rooted in geography and geology and incorporating many aspects of hydrology, climatology, and pedology – is well served by the forum provided by CATENA. Much progress has been accomplished in fluvial geomorphology during those twenty years, but the highly complex and delicate relationships between channel processes and catchment controls still raise intriguing problems. Concepts dealing with thresholds and systems, and modern tools such as remote sensing and sophisticated tracing, have not decisively resolved the simple but elusive dual problem: how does the catchment shape the stream channel and valley to its form, and why? And: how does the channel transmit its influence upstream in order to make the catchment what it is?

Partial solutions, in a regional or thematic sense, are common and important. In addition to contributing a building block to the study of fluvial geomorphology, they also produce a number of new questions. The consequent proliferation of research topics characterises this collection of papers. The basic tool of geomorphological interpretation – the magnitude, frequency, and mechanism of sediment and water conveyance – is a prime focus of interest. Increasingly important in this context in recent years is the role of human interference natural fluviomorphic process systems. Effects of drainage ditching, transport of pesticides absorbed in fluvial sediment, and the flushing of nutrients are some of the Man-conditioned aspects mentioned in this volume. Other contributions deal with the intricate balance, especially in extreme climatic zones, between physical process generalities and macroregional morphoclimatic influences.

The contributions of PICKUP and of PICKUP & WARNER represent two of the very few detailed quantitative geomorphological analyses of very humid tropical catchments. The 8 to 10 m mean annual rainfall in the equatorial mountain areas studied combines with effective landsliding to produce extremely high denudation rates. However, many aspects of channel behaviour are similar to those of temperate rivers. Particularly interesting are the relationships derived between channel characteristics, perimeter sediment and bedload transport.

Several small ephemeral and intermittent streams in Ohio studied by THARP, although variable in catchment area and in peak discharge, have a similar competence; while sorting increases downstream, the coarsest sizes tend to remain constant. Sorting of fluvial sediment, though on a much longer time scale and in an arid climate, also plays an important role in the contribution by MAYER, GERSON & BULL. They find that modern channel sediment size exhibits the most rapid downstream decrease in mean particle size, while Pleistocene deposits show the least rapid decrease and are consistently finer than younger deposits. The difference is attributed to climatic change and a predictive model thereto is presented.

HASSAN, SCHICK & LARONNE describe a new method for the magnetic tracing of large bedload particles capable of detecting tagged particles redeposited by floods up to several decimetres below the channel bed surface. Their method may considerably enhance the value of numerous experiments with painted pebbles, previously reported or currently in progress. Suspended sediment is the subject of the paper by CARLING. He experiments with sampling gravel-bedded flashy streams by two methods, and concludes that pump-sampling and 'bucket' sampling show significant differences only for very high discharges. Suspended sediment concentration is also dealt with by GURNELL & FENN, but in a proglacial environment – a climatic zone about which our knowledge is largely deficient. They find some correspondence between 'englacial' and 'subglacial' flow components and the total suspended sediment concentration.

The effects of human interference by ditching in a forest catchment on sediment concentration and sediment yield is discussed by BURT, DONOHOE & VANN. A local reservoir afforded an opportunity to monitor in detail the influence of these drainage operations on the sediment concentration which increased dramatically, and, after several months, gradually recovered due to revegetation. TERNAN & MURGATROYD analyse sediment concentrations and specific conductance in a humid, forest and marsh environment. Permanent vegetation dams are found to influence sediment concentration directly through filtration and indirectly through changes in water depth and velocity. Changes in specific conductance are influenced by marsh inputs as well as by the addition of area of coniferous forest. The relationship between quality of water and fluvial sediment characteristics is dealt with by HERRMANN, THOMAS & HÜBNER, who analyse the regional pattern of estuarine transport processes. They conclude that high pesticide concentrations are correlated with high concentrations of suspended sediment. Hydrodynamic rather than physicochemical factors influence the regional distribution in the estuary, and the effect of brooklets draining intensively cultivated land is quite evident.

Asher P. Schick

CONTENTS

G. PICKUP
    *GEOMORPHOLOGY OF TROPICAL RIVERS
    I. LANDFORMS, HYDROLOGY AND SEDIMENTATION IN THE FLY AND LOWER PURARI, PAPUA NEW GUINEA*

G. PICKUP & R. F. WARNER
    *GEOMORPHOLOGY OF TROPICAL RIVERS
    II. CHANNEL ADJUSTMENT TO SEDIMENT LOAD AND DISCHARGE IN THE FLY AND LOWER PURARI, PAPUA NEW GUINEA*

P. A. CARLING
    COMPARISON OF SUSPENDED SEDIMENT RATING CURVES OBTAINED USING TWO SAMPLING METHODS

J. L. TERNAN & A. L. MURGATROYD
    THE ROLE OF VEGETATION IN BASEFLOW SUSPENDED SEDIMENT AND SPECIFIC CONDUCTANCE IN GRANITE CATCHMENTS, S. W. ENGLAND

T. P. BURT, M. A. DONOHOE & A. R. VANN
    CHANGES IN THE SEDIMENT YIELD OF A SMALL UPLAND CATCHMENT FOLLOWING A PRE-AFFORESTATION DITCHING

R. HERRMANN, W. THOMAS & D. HÜBNER
    ESTUARINE TRANSPORT PROCESSES OF POLYCHLORINATED BIPHENYLS AND ORGANOCHLORINE PESTICIDES – EXE ESTUARY, DEVON

W. SEILER
    MORPHODYNAMISCHE PROZESSE IN ZWEI KLEINEN EINZUGSGEBIETEN IM OBERLAUF DER ERGOLZ – AUSGELÖST DURCH DEN STARKREGEN VOM 29. JULI 1980

A. M. GURNELL & C. R. FENN
    FLOW SEPARATION, SEDIMENT SOURCE AREAS AND SUSPENDED SEDIMENT TRANSPORT IN A PRO-GLACIAL STREAM

T. M. THARP
    SEDIMENT CHARACTERISTICS AND STREAM COMPETENCE IN EPHEMERAL AND INTERMITTENT STREAMS, FAIRBORN, OHIO

L. MAYER, R. GERSON & W. B. BULL
    ALLUVIAL GRAVEL PRODUCTION AND DEPOSITION – A USEFUL INDICATOR OF QUATERNARY CLIMATIC CHANGES IN DESERT (A CASE STUDY IN SOUTHWESTERN ARIZONA)

M. HASSAN, A. P. SCHICK & J. B. LARONNE
    THE RECOVERY OF FLOOD-DISPERSED COARSE SEDIMENT PARTICLES – A THREE-DIMENSIONAL MAGNETIC TRACING METHOD

RILLS ON BADLAND SLOPES: A PHYSICO-CHEMICALLY CONTROLLED PHENOMENON

A.C. **Imeson** & J.M. **Verstraten**, Amsterdam

Summary

The dynamic response of badland regolith material to wetting by rainfall is described for locations in SE-Spain. This response reflects physico-chemical conditions in the regolith. Where rills occur specific responses to wetting are found. Based on these findings a model of rill initiation is put forward for badland slopes in which rills act as drains. Since a rill by draining the adjacent regolith produces a more stable regolith in its immediate vicinity, the parallel arrangement of rills can be explained. An important condition influencing rill development is that of a high macroporosity produced by shrinking and swelling. It is suggested that this might be analogous to the macroporosity produced by ploughing on cultivated land.

Resumen

La respuesta del manto rocoso a tierras malas, se refleja conditiones fysico-quimico de la regolita. Uno territorio con carcavas tienne respuestas typicas a aqua. Fundado en la resulta de investigationes , uno modele hace desarrollado que explica el evolucion de carcavas joven en tierras malas. Para el evolucion de las carcavas, la regolita adquire un drenaje mejor y por eso uno suelo masestable. El patron de carcavas paralelo esta la resulta de esto evolucion. Una condition inportanta para el evolucion de las carcavas esta la macroporosidad del suelo para contrar y hinchar.

1 Introduction

Erosional processes have originally been explained by invoking mechanical forces. However, in some cases and especially in semi-arid areas or on cohesive materials, the way in which materials behave under rainfall is influenced by physico-chemical processes. This behaviour in turn influences the hydraulic and hydrologic characteristics of the material being eroded. An explanation of erosional phenomena in terms of mechanical forces alone results in an incomplete visualisation of the erosional processes involved. In this paper it is suggested that the responses of different badland regolith materials to wetting by rainfall might account for the distribution and symmetry of rills observed in the field. A model of rill development is described, based on inferences made from the association of particular material properties and rill occurrence, and on observations and measurements made during field experiments with rainfall simulators. The degree to which this model can be applied to rills which develop on cultivated land is considered briefly. This paper is based on the results of research programmes being undertaken in areas of badlands along the Rio Fardes and Rio Guadahortuna in southeast Spain, described in detail by GERITS (1987) and in the Dinosaur Park badlands along the Red Deer River in Alberta, Canada (BRYAN 1982). It forms an extension of an earlier paper (GERITS et al. 1987) in which an association between certain dynamic regolith properties and rill occurrence was observed in the same badland areas.

The development of rill systems was explained by HORTON (1945) in terms of the erosive force of overland flow. HORTON argued that as the volume of overland flow increases with slope length, so does its depth and velocity. At a critical velocity and distance from a divide rills would develop and subsequently evolve into larger channels by cross-grading and micro-piracy. Most recent research on rill erosion has concentrated on hydraulic aspects of rill initiation and enlargement, implicitly assuming that rills originate due to the entrainment of particles by overland flow, once a particular set of threshold hydraulic conditions has been reached. This work has recently been reviewed by DE PLOEY (1984) and GOVERS (1985, 1987). These authors found that for non-cohesive loess soils in Belgium, rills formed when a threshold shear velocity of between 3 and 3.5 cm/s was exceeded; at this point non-selective erosion of loess could occur (GOVERS 1986). The findings of these authors were based on both field and laboratory results.

Most literature on rill erosion refers to rills developed on cultivated land. Emphasis has been given to the distinction between interrill and rill erosion (MEYER et al. 1972) because this can be used to facilitate the modelling of soil loss. Rills on cultivated land and badland rill systems are morphologically similar and perhaps formed by similar processes. An important difference between cultivation and badland rill systems, however, is that by definition rills on agricultural land are formed between successive tillage operations. As they grow they supply large amounts of sediment to downslope areas. Badland rill systems are sometimes obliterated by seasonally occurring processes and grow again each year. However, badland rill systems may be relatively old and exist in a state of dynamic equilibrium. Under such conditions the occurrence of rills might not signify a high rate of soil loss.

One problem of investigating rill systems is that the conditions existing before the process of rill erosion began, may no longer exist once the rill has formed. Measuring the hydraulic conditions on rill and interrill areas is not necessarily helpful in explaining why and when rills develop. The shear strength and hydraulic conductivity are dynamic properties of the regolith and it is important to establish these parameters at the time of rill initiation.

2 The Response of Badland Materials to Rainfall

When soil or regolith materials are moistened by rainfall, the way in which they respond is controlled by a number of phyico-chemical processes. These processes are themselves controlled by the mineralogical, chemical and textural

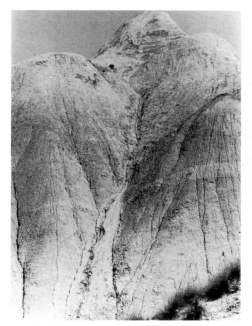

Photo 1: *Rilled slopes at the Cortijo site, Rio Fardes, Spain.*

Photo 2: *Rilled slopes at the Cortijo site, Rio Fardes, Spain.*

properties of the material. For the badlands described in this paper, these basic regolith properties are reported by BRYAN (1984) and GERITS (1988). The types of response which occur and how these can be measured are discussed in GERITS et al. (1987). In brief, the most important responses are swelling and shrinking (volumetric changes), slaking and crust development, flow and the dispersion of clay minerals. During a rainfall event, as the regolith absorbs moisture, the above mentioned responses or processes exert a large influence on the surface and internal drainage of the regolith. Interactions between water and regolith result in a dynamic solution chemistry. This is important because it influences the amount and rate of swelling, the dispersion of clay and amount of sediment released into suspension, and the erodibility and erosivity of respectively the regolith and flowing water.

During an investigation reported elsewhere (GERITS et al. 1987), it was noted that rills occurred in regolith materials having particular combinations of properties. Where rills occurred, the regolith had a relatively large capacity to shrink and swell, a high slaking index and a high liquid limit. These properties could not have extremely high values, otherwise erosional processes other than rill erosion occurred. Based on this association it was inferred that rill development was possible when the regolith had a high macroporosity, or crack density, when infiltration was discontinuous, and when the regolith could resist a too rapid tendency to distortion by flow. It was further postulated that if rills were considered as drains, simple steady-state drainage equations could be applied to rills in a way which explained the parallel symmetry of shoestring rills, their distribution in the investigated badland area and the significance of the regolith properties with which they are associated. Before applying these findings to a consideration of how rills might evolve, it is useful to briefly consider the most common characteristics of badland regolith profiles and the way in which they disperse, slake and swell.

3 Characteristics of Badland Regolith Material

Badland regolith materials developed on shales and marls in the two badland areas have similar morphological characteristics (GERITS et al. 1987). The regolith can usually be divided into the following layers:

1. surface crust, 1–2 cm thick, broken by tension or shrinkage cracks.

2. 10–25 cm thick subcrust, with a relatively large number of macropores when dry; when moist often compact.

3. transitional layer, resembling above horizon except that partially weathered shards from underlying layer are recognisable.

4. subangular shards into which the weathered parent layer breaks up when exposed to atmosphere conditions.

Detailed descriptions of regolith profiles are given in the literature cited. It is considered important that horizons 2 and 3 are composed of distinct peds, usually several cm in diameter, separated by macropores which often contain material deposited by subsurface flow. The crust may take on a number of forms, having not only a different morphology but also different amounts of water soluble salts (GERITS 1987).

The types of rill referred to in this paper are illustrated in photos 1 and 2. Photos 3A and 3B and 4A and 4B respectively show sections across rill systems at the Cortijo site, along the Rio Fardes in Spain and at a site in the Dinosaur Park badlands in Canada.

It should be noted that the shard-regolith boundary is at the surface under the rill as sketched schematically in fig.1. Unlike in the schematic figure, however, this boundary is closer to the surface at the Dinosaur Park badlands site.

4 Methods Used to Assess Response to Wetting

The way in which samples of the different regolith horizons reacted to wetting was evaluated using procedures which have been reported in detail elsewhere (GERITS 1987, GERITS et al. 1987, IMESON 1986, IMESON & OOSTWOUD WIJDENES 1988). The volumetric changes which occurred when moistened natural aggregates were dried was measured using Seran resin. Increases in volume during moistening were measured photographically. The results were used to calculate the shrinkage limit (Ws), the COLE 25 value, the swelling limit (Swl) and shrinkage ratio (S.R.). The consistency parameters: Liquid limit (LL), Plastic limit (P.L.), consistency index of DE PLOEY (C5-

Rills on Badland Slopes

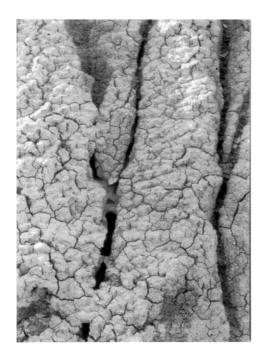

Photo 3: *A and B: Section across rilled slope, Cortijo site, Rio Fardes, Spain.*

Photo 4: *A and B: Section across rilled slope. Dinosaur Park Badlands, Canada, before and after excavation.*

10), activity (A) and the plasticity index (P.I.), were determined using standard procedures (SINGH 1967, DE PLOEY & MUCHER 1981). Several dispersion indices were calculated as well as the sodium adsorption ratio (SARp) and cation composition of the regolith solution and runoff.

5 Rills as Drains

It was suggested that association of rills with particular regolith properties could be explained if rills functioned as drains (GERITS et al. 1987). Simple equations describing drain discharge as a function of spacing, and hydraulic conductivity such as the equation, taken from WESSELING (1973):

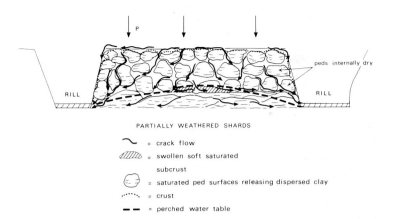

Fig. 1: *Schematic diagram of water movement in rills in badland regolith materials.*

$$Q = \frac{(8 \cdot Kb \cdot D \cdot h) + (4 \cdot Ka \cdot h^2)}{L^2}$$

where Q is the drain discharge, K is the hydraulic conductivity above (Ka) or below (Kb) drain level, D is the thickness of the aquifer below drain level, h is the hydraulic head for subsurface flow into drains, and L is drain spacing. This could be applied to a regolith profile such as that in fig.1, where rills extend into a regolith drained by macropores (cracks).

Taking each property in turn, the dynamic effect of these on the development of rills as drains is thought to be as follows:

6 Swelling and Shrinking

Shrinking and swelling are important in creating macropores and in enabling infiltration to be discontinuous. Infiltrating water short-circuits the bulk of regolith in a way analogous to that described by BOUMA et al. (1981).

Shrinking and swelling have this effect on the regolith because the decrease in bulk density which occurs when shards swell, is irreversible. Infiltrating rain or a humid regolith atmosphere causes an upward expansion to the regolith and the production of tension cracks. The potential increase in volume may be considerable (IMESON 1986). When this swollen material later dries, and if the volumetric shrinkage is large, the shrinkage cracks that develop can occupy at least 20 percent of the volume, resulting in a material having a very low (0.8–1.2) bulk density. The penetration of water into a ped is very slow because the swelling of the outer surface causes a very large decrease in hydraulic conductivity. Too much shrinkling and swelling, especially on steep slopes, produces a loose popcorn-like material which is too mobile or dynamic to enable rills to persist or micropipes to develop into larger features. Too little swelling can have several consequences. Firstly the amount of infiltration is inadequately concentrated, so that uniform infiltration occurs with a distinct wetting front. The volume of subsurface flow may be adequate to produce an occasional pipe but too little to enable rills to form.

Photo 5: *The Dehasas badland slopes in which rills were not observed.*

7 Consistency

Rill occurrence appears to be associated with a relatively high liquid limit and with consistency parameters that are considered to be related to slaking. A high liquid limit is thought to be important because it reflects the ability of material to resist displacement by shallow sliding. At sites in SE Spain, shallow sliding occurs very frequently where the liquid limit is low and the regolith too unstable for rills or continuous pipe systems to suvive (photo 5). The liquid limit is not necessarily of general importance with respect to rill occurrence. At the sites studied it is related to a threshold between two exclusive sets of erosional processes; one dominated by flow through macropores which are in material that offers s relatively high resistance to liquefaction; the other in material in which flow processes occur at such a low threshold moisture content that subsurface micropipes are unable to form or survive.

The C5-10 index of DE PLOEY & MÜCHER (1981) indicates that at high moisture contents instability is increased by the addition of relatively less moisture at the sites where rills occur. Low values of the C5-10 index have been shown to be associated with slaking by DE PLOEY (1981). The slaking of ped faces could have the same effect on promoting discontinuous water penetration as swelling.

8 Dispersion

A comparison of the SAR and EC values of both saturated paste extracts and runoff, at the sites where rills are either present or absent, indicated that at the rilled sites clay is more likely to be dispersed. This has been corroborated by flocculation expriments (GERITS 1987). GERITS (1987b) has shown that due to the effects of the chemistry of the water soluble salts on swelling and dispersion, the critical shear stress for material from the rilled sites to be entrained is lower that at the non-rilled sites. Very high concentrations of sediment in the more discontinuous runoff at the rilled site may be expected to occur more frequently.

9 Discussion

The general effect of the material properties at the rilled and non-rilled sites may be summarised as follows. At the non-rilled sites water penetration is relatively uniform and somewhat deep. Sheetflow (studied on video recordings made during experiments: GERITS 1988) is present over large areas but because of its uniform character entrains relatively lower amounts of sediment. Relatively larger amounts of precipitation are needed to generate runoff. However, prolonged rainfall can lead to the flow or sliding of the wetted surface regolith to channel areas. At the rilled sites water penetration is variable in depth due to the effects of swelling and shrinking, and runoff occurs at lower threshold rainfall intensities. Runoff is concentrated by crack systems and voids and rill or pipe systems receiving this water rapidly produce flow. The rills or pipes develop in material which releases dispersed clay but which retains its cohesion sufficiently to resist deformation.

The material properties associated with rilled or non-rilled slopes may not necessarily be associated with rills alsewhere, as they are probably specific to steep badland lopes. Nevertheless, several conclusions can be drawn which are of wider significane. The main conclusion is that when soils swell and seal and contain cracks, flow can be concentrated at low rainfall intensities in a way that leads to rill formation. During the last year the authors have examined vry many rilled sites on road embankments in northeast Spain and elsewhere (photo 6) and in virtually all cases rills are confined to material which has developed shrinkage cracks and which is often dispersive. Particularly well developed B or BC horizons of Pleistocene paleosols or soils having a duplex character are subject to rilling. When the role of shrinking and swelling in concentrating water is apparent, it is easy to explain why rills begin at the top of embankments on slopes without a belt of no erosion. It is also possible to envisage fluting as an extreme form of rill development.

A schematic model of rill development appropriate to badland slopes or road embankments is shown in fig.1 and 2. The essential element of the model is the relationship between the resistance of the regolith to displacement by mass-movement (liquefaction and flow) or entrainment by crack flow and moisture content. A rill or pipe is assumed to develop in the regolith where crackflow concentrates water to a level where a critical moisture content is reached, necessary for a sudden increase in erosion to occur. Once formed, the rill drains the adjacent regolith so that this critical

Photo 6: *Rills comparable to those on badlands in a road cutting near Gerona, NE Spain.*

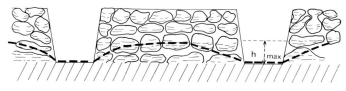

K = crack density
L = distance between rills
h_{max} = max possible elevation of perched water table
— — = perched water table

Fig. 2: *The critical elements of the drainage model of rill initiation. Above h_{max}, the perched water table enters unstable material (weathered shards) in a moist state.*

moisture content is not reached until a critical distance. This distance will depend on the efficiency of the crackflow in draining the regolith and preventing the critical moisture content from being reached. In fig.3, a hypothetical situation is shown for a case in which displacement occurs by liquefaction or flow when the weathered regolith above a shard layer is moistened long enough for it to swell into a mobile mass. In this example it is assumed that the critical conditions is the occurrence of a positive pore pressure, in he subcrust material, although any other critical condition related to drainage would have the same

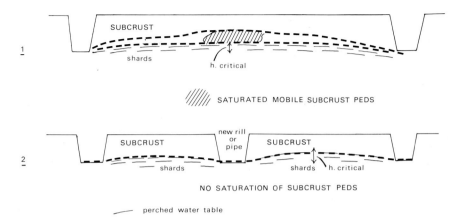

Fig. 3: *Schematic diagram indicating how insufficient drainage of the regolith (1) can lead to the development of an unstable zone. In this zone liquefaction or erosion can result in a pipe or rill developing and the perched water table being lowered into material that is stable when saturated (shards).*

effect. Rills will develop parallel to one another at a distance sufficiently close to prevent the critical condition from occurring. The most important parameters controlling rill spacing are in this case the hydraulic conductivity and slope angle. The steeper the slope, the lower the threshold moisture content will need to be for displacement and the closer rills will need to be to drain the area between time. The less well developed crack systems are (low hydraulic conductivity through macropores), the shorter the distance between rills will need to be to drain the slope to a moisture content below that required for erosion.

The postulated relationship between rill spacing, slope and hydraulic conductivity agrees with observation. It is necessary, however, to obtain field parameters to enable a simulation model of rill formation to be developed. Current research is being undertaken to examine how the shear strength of material containing rills varies during runoff events.

Current research is also being undertaken to test the applicability of the hypotesized model of rill initiation to cultivated soils. On cultivaed soils discontinuous infiltration can occur due to tillage. It is quite possible that where soils are sensitive to slaking, tillage enables the rapid development of saturated zones in topographic depressions where the shear resistance of the soil (often colluvium) is lowered sufficiently for the overland flow, generated by the saturated flow between the peds, to develop channels. The occurrence of rills under such conditions is however less likely to be dependent on physico-chemical soil characteristics than is the case for badland rills.

Acknowledgement

We wish to thank J. Gerits, R.B. Bryan, F.J.P.M. Kwaad and A. Yair for discussions contributing to this paper. Mrs. M.C.G. Keijzer-v.d.Lubbe is thanked for preparing the manuscript, Mr. C. Snabilié for the illustrations.

References

BOUMA, J., DEKKER, L.W. & MUILWIJK, C.J. (1981): A field method for measuring short-circuiting in clay soils. J. Hydrology 52, 347–354.

BRYAN, R.B. & CAMPBELL, I.A. (1982): Surface flow and erosional processes in semi-arid micro-scale channels and drainage basins. I.A.S.H. Hydr. Publ. 137, 123–133.

BRYAN, R.B., IMESON, A.C. & CAMPBELL, I.A. (1984): Solute release and sediment entrainment on microcatchments in the Dinosaur Badlands, Alberta, Canada. Journal of Hydrology, 71, 79–106.

DE PLOEY, J. (1984): Hydraulics of runoff and loess loam deposition. Earth Surface Processes and Landforms, 9, 525–531.

DE PLOEY, J. & MÜCHER, H.J. (1981): A consistency index and rainwash mechanisms on Belgian loamy soils. Earth Surface Processes, 6, 319–330.

GERITS, J.J.P. (1988): Physico-chemical thresholds for sediment entrainment and transport. University of Amsterdam, Ph.D., in preparation.

GERITS, J.J.P., IMESON, A.C., VERSTRATEN, J.M. & BRYAN, R.B. (1987): Rill development and badland regolith properties. CATENA SUPPLEMENT 8, 141–160.

GOVERS, G. (1985): Selectivity and transport capacity in thin flow in relation to rill erosion. CATENA, 12, 35–49.

GOVERS, G. (1987): Spatial and temporal variability in rill development processes at the Huldenberg Experimental Site. CATENA SUPPLEMENT 8, 17–34.

HORTON, R.B. (1945): Erosional development of streams and their drainage basins: hydrophysical approach to quantitative morphology. Bulletin Geol. Soc. of America, 56, 275–370.

IMESON, A.C. (1986): Investigating volumetric changes in clayey soils related to subsurface water movement and piping. Zeitschr. für Geomorphlogie, Suppl. Bd. 59, 115–130.

IMESON, A.C. & OOSTWOUD-WIJDENS, D. (1988): Volume changes of badland regolith profiles. Manuscript.

MEYER, L.D., FOSTER, G.R. & ROMKENS, M.J.M. (1972): Source of soil eroded by water from upland slopes. In: ARS Publ. ARS-5-40, Present and Prospective Technology for predicting sediments yields and sources, 177–189.

SINGH, A. (1967): Soil Engineering in Theory and Practice. Asia Publishing House, London, 805 pp.

WESSELING, J. (1973): Drainage Principles and Applications. Vol. 2, Theories of Field Drainage and Watershed Runoff, ILRI, Wageningen.

Address of authors:
A.C. Imeson and J.M. Verstraten
Laboratory of Physical Geography and Soil Science
University of Amsterdam
Dapperstraat 115
1093 BS Amsterdam, The Netherlands

MEASUREMENT AND ASSESSMENT OF SOIL LOSS IN RWANDA

L.A. **Lewis**, Worcester

Summary

During the 1983–84 agricultural year, data were collected at 100 field sites to assess the amount and spatial distribution of soil loss in Rwanda. Erosivity, erodibility, topography, crop cover and conservation, the factors determining erosional rates on agricultural lands were measured and evaluated. These results were used to calibrate the USLE for the country's agro-ecologic conditions. Correlations between the measured and estimated USLE values were statistically significant. This permits soil loss to be estimated for the 20.000 fields comprising the complete national agricultural survey.

The highest soil loss recorded in either of Rwanda's two growing seasons during the one year field period was 17 t/ha. At the préfecture (regional) scale, average annual soil losses range from 7.1 to 0.4 t/ha. Overall soil losses are low in Rwanda compared to other countries having similar environmental conditions. The widespread cultivation of bananas and the prevailing agricultural practice of intercropping provide good groundcover throughout the rainy season.

The importance of groundcover in minimizing erosion on Rwanda's agricultural lands is substantiated by examining the relation of crop types and conservation practices on soil loss. Perennial crops result in low soil losses compared to seasonal crops. And annuals which provide poor groundcover during the first few months of each rainy season, such as maize, are associated with higher erosion than annuals, such as beans, which provide a better groundcover earlier in the growing season. With increasing cover, the effectiveness of raindrop impact in initiating erosional processes is reduced. Finally, mulching contrasted to engineering methods is found to be the most effective conservation practice even on slopes exceeding 30 degrees.

Résumé

Afin d'estimer le volume et la distribution géographique de perte en terre au Rwanda, des données furent recueillies sur 100 installations pendant l'année agricole 1983–84. Le susceptibilite des sols à l'érosion, l'aggressivité de la pluie, la topographie, le couverture végétale du sol, la méthode de lutte contre l'érosion, et autres déterminants de l'érosion du sol furent mesurés et evalués. Ces résultats furent utilisés dans la calibration de l'équation universelle de perte en terre pour les conditions agro-écologiques du Rwanda. Les corrélations entre les valeurs mesurées et estimées

ISSN 0722-0723
ISBN 3-923381-12-3
©1988 by CATENA VERLAG,
D–3302 Cremlingen-Destedt, W. Germany
3-923381-12-3/88/5011851/US$ 2.00 + 0.25

de l'équation universelle se sont avérées significatives, permettant l'estimation de perte en terre sur les 20.000 champs étudiés dans l'Enquête nationale agricole.

La perte en sol le plus élevée enregistrée dans un champ pendant une des deux saisons de l'année agricole a été de 17 tonnes a l'hectare. Au niveau régional, la moyenne annuelle de perte en sol varie de 7.1 a 0.4 tonnes a l'hectare. Comparée aux autres pays ayant les mêmes conditions écologiques, l'érosion du sol au Rwanda et peu élevée. La culture repondue de la banane et la pratique de plusieurs cultures dans le même champs fournissent une bonne couverture du sol pendant les deux saisons pluvieuses.

L'importance de la couverture du sol pour minimiser l'érosion sur les exploitations Rwandaise est confirmée par une analyse de l'influence des types de cultures et les méthodes de lutte contre l'érosion sur la perte en sol. Les cultures pérennes permettent moins d'érosion du sol par rapport aux cultures annuelles. Les cultures annuelles, tel que le mais, qui pourvoient une faible couverture de sol pendant les premiers mois de chaque saison de pluie, sont associées à un niveau d'érosion plus élevé que les cultures annuelles, tel que le haricot, qui pourvoit une couverture du sol très tot dans la saison. L'amélioration de la couverture du sol réduit l'impact de la pluie dans le processus d'érosion. En dernier lieu, par rapport aux autres méthodes de lutte contre l'érosion employées au Rwanda (terrasses, fosses, haies, etc.) le paillage est considéré comme la plus effective forme de conservation, même sur les pertes dépassant 30 dégrés.

Resumen

Durante el año agricola 1983–84 se recogieron datos en 100 estaciones con el fin de estimar el valor y la distribución espacial de la pérdida de suelo en Rwanda. Se tomaron mediciones y se evaluaron los factores que determinan las tasas de erosion en las tierras agricolas, tales como erosividad, erodibilidad, topografia, tipo de cultivo y prácticas de conservación. Estos resultados se utilizaron para calibrar la EQUACION UNIVERSAL DE PERDIDA DE SUELOS (USLE) en las condiciones agro-ecológicas del país. Las correlaciones entre los valores medidos y los estimados fueron estadísticamente significatios. Esto permite evaluar la pérdida de suelos para los 20.000 campos que comprende el total del territorio nacional.

La mayor pérdida de suelo registrada en cualquiera de las dos épocas de crecimiento de las plantas fué de 17 t/ha. A escala regional las pérdidas anuales medias de suelo oscilan entre 7.1 y 0.4 t/ha. En conjunto las pérdidas de suelo en Rwanda son bajas si se comparan con las de otros paises de condiciones ambientales similares. El extenso cultivo de bananas y la práctica agrícola preponderante de intercalar cosechas proporiona un buen recubrimiento del suelo durante todas las estaciones lluviosas.

La importancia del recubriminto del suelo en la minimización de la erosión en las tierras agricolas de Rwanda se hace patente al examinar la relación entre tipos de cultivos y prácticas de conservación. Los cultivos contínuos presentan pérdidas de suelo bajas comparadas con las de los cultivos estacionales, aunque no en el caso de plantas con un recubrimiento débil del suelo durante los primeros meses de la estación lluviosa,

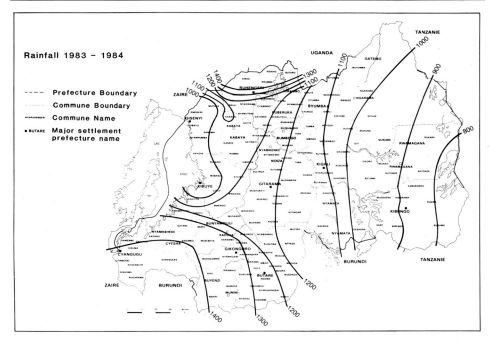

Fig. 1: *Rainfall during 1983–84 agricultural season.*

como el maíz por ejemplo, lo cual no sucede en el caso de los frijoles, que proporcionan un mejor recubrimiento. El efecto del impacto de las gotas de lluvia en el inicio de los procesos de erosión queda sensiblemente reducido con el incremento de la cobertura del suelo por las plantas. Finalmente, el recubrimiento del suelo con una capa de paja, incluso en vertientes de más de 30°, es mucho más efectivo como método de conservación que las construcciones de ingeniería.

1 Introduction

Rwanda's population density of 191.3/km^3 (BUREAU NATIONAL DE RECENSEMENT 1982, 15) makes it one of the most densely populated agricultural countries in the world. Approximately 90% of its inhabitants are directly engaged in agriculture (MINISTERE DE L'AGRICULTURE ET DE L'ELEVAGE 1983b, 1) and its population density per cultivated land is about 400/km^2 (WORLD BANK 1983, 2). Considering both its topographic and soil characteristics, 30.8% of its cultivated lands are classified as good, 48.9% average, and 20.3% poor (MINISTERE DE L'AGRICULTURE ET DE L'ELEVAGE 1983a, 39).

Generally, annual precipitation reflects elevation. This results in the higher western portion of the country receiving greater precipitation than the eastern half (fig.1). Because of the relatively low rainfall in the east, the majority of the best rainfed agricultural lands are found in the west inspite of the prevailing hilly/mountainous terrain. It appears that all the lands that should be cultivated within the more humid portions of the nation are currently farmed. In fact, land scarcity in the better agricultural areas has resulted in recent rural migration toward the drier lower eastern lands.

Because of their steep slopes and local rainfall characteristics, a significant percentage of Rwanda's farmlands have a high potential for excessive soil erosion if not properly managed. The magnitude and intensity of the tropical showers require a good groundcover to minimize the erosional potential of raindrop impact. Because this area of Africa has two definite rainy seasons, management practices need to counter the existence of bare ground at the onset of two growing seasons if soil losses are to be minimized on the agricultural lands. In many nations having similar physical environmental settings and like Rwanda also experiencing increases in rural population, data indicate that rates of soil erosion are increasing and long-term agricultural potential is decreasing (LEWIS & COFFEY 1985, LEWIS 1985, REPETTO 1986).

Prior to the first national agricultural survey, little erosional data existed for Rwanda. However, experimental studies in selected areas indicated that soil loss was excessive (WASSMER 1981, 65). These high losses could threaten the long-term agricultural potential of the country if the rates for these specific areas were indicative of general prevailing conditions. To assess the agricultural erosional situation throughout Rwanda, 100 field sites located on fields within active farms were established throughout the country for a one-year period (1983–84). This data collection was undertaken as a component in Rwanda's first national agricultural survey.

2 Study Sites

Based on a preliminary investigation, 100 field sites, from eight to eleven in each préfecture, were established to obtain measurements of soil loss (fig.1). The specific fields chosen for this study were selected from the 10.000 sample fields comprising the total sample population of the agricultural survey for each growing season. Both the sample size and the areal distribution of the field sites permitted a wide range of the typical agricultural settings existing in Rwanda to be monitored during the one-year study of soil loss. The major environmental settings and the primary crops and crop mixes are included within the soil loss sample.

Range in degrees	First season Number	Second season Number
0.0-4.9	11	7
5.0-9.9	22	27
10.0-14.9	28	25
15.0-19.9	16	14
20.0-24.9	12	13
25.0-29.9	8	6
30.0-35.0	2	5
Total number of fields	99	97

Tab. 1: *Slope angles of field sites.*

The hilly nature of Rwanda's terrain is reflected in the range and magnitude of the sample fields' slopes. Slope angles range from less than 1 degree (1%) to 35 degrees (70%) (tab.1). The average slope for the first growing season is 12.9 degrees (23%); for the second season it is 13.6 degrees (24%). Bananas, the dominant crop, grew in 31% of the sample fields as either the primary or secondary crop. In total, 47 crop mixes were sampled over the two growing seasons.

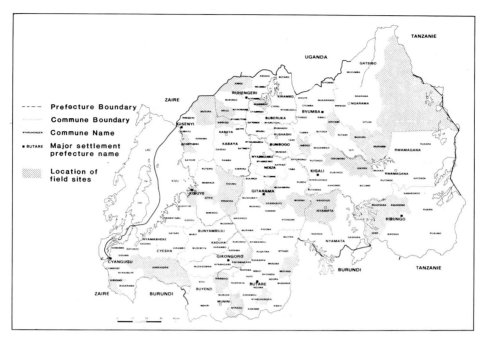

Fig. 2: *Location of field sites.*

Besides bananas, the major crops were beans, sorghum, manioc, corn (maize), and sweet potatoes.

Total annual rainfall during the 1983–84 study period was generally about 80% of the long-term average at the majority of stations throughout the country (DIVISION DE CLIMATOLOGIE 1985, 1986). These rainfall amounts fall well within the 'normal' expected annual precipitation range. Thus the study year is considered to be typical, albeit slightly dry. In the humid northwest (Gisenyi-Ruhengeri), the annual rainfall was 1400 mm; in the drier southeast (Gashora-Kibungo) annual rainfall was 850 mm during the study period.

3 Data Collection

The immediate aim of the data collected in this study is to provide the necessary information to accurately assess and understand the soil loss situation throughout Rwanda. However, equally important is one of the long-term goals of the agricultural survey, to identify the variables that need to be incorporated in future agricultural surveys to provide agricultural planners with sufficient information to maintain long-term agricultural productivity and protecting the environment from degradation caused by soil erosion.

The Universal Soil Loss Equation (USLE) (WISCHMEIER & SMITH 1978) forms the framework for this study's inquiry. According to this equation, the six critical factors that deter-

Photo 1: *Soil trap installed in a newly prepared field prior to the beginning of the rainy season.*

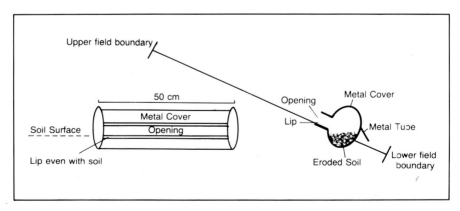

Fig. 3: *Diagram of soil trap and generalized location of traps in fields.*

mine soil loss are: rainfall (erosivity), erodibility, slope length, slope steepness, crop cover and management, and conservation. In this study with the exception of rainfall and erodibility each of the other factors were collected directly in each field, for each rainy season. Rainfall and erodibility had to be inferred for each site. Erosivity was computed using the rainfall data collected at the nearest meteorological station; erodibility was based on the area's parent material, the local bedrock. Soil loss data were collected by installing soil traps at the bottom of each of the 100 sample fields (fig.2, photo 1). The soil traps are slight modifications of the sediment troughs used in Kenya (LEWIS 1985). Each trap is 50 cm wide with a diameter of 130 mm. The traps have an approximate capacity to hold about 2 kilos of soil.

Soil loss samples were collected on a weekly basis rather than the usual practice of data collection after each rain event. Each week the soil collected in the soil traps was transferred into plastic bags. These bags of soil were oven-dried and weighed according to standard practices. An independent evaluation of the reliability and validity of both the traps and weekly collection of data determined that this method, developed for use in Kenya (LEWIS 1985), produces comparable results to those obtained from standard runoff plots at significantly lower costs (FELDMAN, FREEMAN, O'ROURKE & McGAHUEY 1985).

For the two growing seasons, only in nine cases were the traps' capacity inadequate to collect all of the soil loss washed downslope during the one week collection interval. In those cases the field data likely underestimate the actual soil loss. However, the weekly soil loss values computed using the full capacity of the traps always results in high erosional values. But because there is an upper limit on computed weekly soil losses, resulting from the limited capacity of the traps, the overall soil loss values of this study are considered to be on the conservative side.

Both slope length and angle were determined by direct field measurements using standard methods. A steel tape, accurate to a tenth of a meter, was used to measure the length of the field perpendicular to the soil trap. Average slope steepness was determined by direct measurement for the slope segment from the upper field boundary to the soil trap. The catchment area for each trap was assumed to be the width of the trap (50 cm) multiplied by the length of the field. Because in almost all cases slopes are uniform for each field, the assumed catchment area approximates the actual situation.

Crop cover for each field was determined in the second month of each growing season. For classification, five categories were used:

1. excellent - complete groundcover;

2. good - irregular bare ground patches exist, but always less than 5 cm in length;

3. average - bare ground exists between crop rows, but less than 15 cm;

4. below average - bare ground up to 30 cm existing between or within crop rows; and

5. poor - bare ground existing that is greater than that in the below average category (tab.2).

Category	Weight (added to C-value)
excellent	0.00
good	0.02
average	0.04
below average	0.07
poor	0.10

Tab. 2: *Groundcover categories and weights assigned to C-values.*

	Crop	C-value		Crop	C-Value
1.	Coffee	0.02	25.	Manioc/cocoyam	0.20
2.	Banana	0.04	26.	Pyrethrum	0.20
3.	Banana/beans	0.10	27.	Sorghum/sweet potato	0.20
4.	Harvested field	0.10	28.	Peanut/beans	0.21
5.	Manioc/banana	0.10	29.	Potato/maize	0.21
6.	Not cultivated	0.10	30.	Manioc/beans	0.22
7.	Pasture	0.10	31.	Manioc/sweet potato	0.22
8.	Soy beans/banana	0.11	32.	Potato	0.22
9.	Beans/banana	0.12	33.	Sweet potato	0.23
10.	Banana/cocoyam	0.13	34.	Eleusine	0.25
11.	Banana/sorghum	0.14	35.	Maize/sweet potato	0.25
12.	Sweet potato/beans	0.14	36.	Manioc	0.26
13.	Maize/banana	0.15	37.	Beans/maize	0.30
14.	Peas	0.15	38.	Maize/beans	0.30
15.	Sweet potato/banana	0.15	39.	Maize/peas	0.31
16.	Sorghum/cocoyam	0.17	40.	Sorghum/manioc	0.31
17.	Sorghum/banana	0.18	41.	Maize/tobacco	0.32
18.	Cocoyam/banana	0.19	42.	Cocoyam	0.35
19.	Beans	0.19	43.	Maize	0.35
20.	Beans/manioc	0.20	44.	Sorghum/maize	0.35
21.	Beans/peanut	0.20	45.	Manioc/quinine	0.38
22.	Beans/peas	0.20	46.	Sorghum	0.40
23.	Beans/potato	0.20	47.	Tobacco	0.45
24.	Beans/sweet potato	0.20			

Tab. 3: *Crop values (C-values).*

Categorie	P-Value	Criteria (conservation practices)
1) Worst	0.35	no evident conservation
2)	0.25	cut-off trench
3)	0.25	terrace
4)	0.20	terrace with cut-off trench
5)	0.15	mulching
6)	0.15	mulching with cut-off trench
7)	0.15	mulching and terrace
8)	0.10	mulching, cut-off trench, and terrace

Tab. 4: *Conservation factor (P-value).*

Préfecture and season		No of fields soil loss >5 t/ha & percentage		Soil loss/field in t/ha/season (to nearest ton)	Average soil loss in the Préfecture (t/ha)
Butare	1	2	20%	1,4,2,0,6,0,14,2,1,2	3.2
	2	0	0%	0,0,0,1,0,1,0,0,1	0.5
Byumba	1	0	0%	0,2,4,3,1,2,1,1,2,0,0	1.4
	2	0	0%	1,1,0,0,1,0,0,0,1	0.4
Cyangugu	1	2	25%	12,2,1,10,2,1,0,1	3.7
	2	0	0%	0,0,0,0,1,0,0,0	0.2
Gikongoro	1	3	30%	13,1,0,1,0,1,2,2,6,17	4.3
	2	3	30%	11,2,0,1,0,5,2,1,6,1	2.8
Gisenyi	1	0	0%	1,1,1,2,0,1,0,0,0,1	0.6
	2	1	11%	1,0,2,2,2,6,0,0,0	1.6
Gitarama	1	2	22%	1,14,0,0,5,1,4,3,1	3.1
	2	0	0%	0,0,0,0,0,0,0,0,1	0.2
Kibungo	1	0	0%	1,0,0,0,0,0,0,0,0	0.2
	2	0	0%	0,0,0,0,0,0,0,0	0.2
Kibuye	1	6	55%	6,5,6,1,3,5,1,0,7,8,4	4.3
	2	0	0%	0,1,0,2,1,3,1,2,1	1.1
Kigali	1	0	0%	1,0,0,1,0,0,0,0,1,3	0.6
	2	0	0%	0,0,0,0,0,0,0,0,0	0.1
Ruhengeri	1	1	10%	0,1,0,2,5,0,0,0,1,0	1.1
	2	0	0%	1,0,1,0,1,0,3,0,1,1	1.1

Tab. 5: *Soil loss by Préfecture 1983–1984.*

In addition to the status of the groundcover, the primary and secondary crops grown in the fields were determined. For the field period, a total of 47 different crop combinations were monitored (tab.3). Within Rwanda, eight different combinations of conservation practices were observed. These ranged from no obvious conservation being practiced to fields that were terraced, mulched, and had cutoff trenches constructed to divert runoff around the fields (tab.4).

4 Results and Discussion

4.1 Measured Soil Loss

In the one year study period only three traps were surreptitiously removed (lost) from the fields. Of the 3.722 (1.782 1st season + 1.940 2nd season) weekly observations, soil loss was measured in 880 of these (49%) during the first season (Nov.–April) and in 562 (29%) during the second season (May–Oct.). For the first season the highest soil loss was 17.0 t/ha observed in Gikongoro (cocoyam); during the second season the greatest loss was 11.2 t/ha (manioc/quinine), again in Gikongoro. As field boundaries in many cases change from season to season, it is not valid to sum the soil loss values collected during the two seasons to determine the annual soil loss from specific sites. Tab.5 summarizes the measured soil loss, by prefecture, for the 1983–84 growing seasons. These data when compared to soil losses, under similar settings in Kenya (LEWIS 1985, 281), are overall exceedingly low.

Given the general thickness and bulk density characteristics of soils throughout the country, this study assumes that a soil loss greater than 5 t/ha will, in most cases, exceed a field's soil tolerance value. Hence productivity in the long-term will likely decline in such areas. Given the goals of this study, the identification of areas where fields have such high soil losses is of paramount importance. For the conditions experienced during the study period, the major soil losses occurred during the first season. While the Préfectures of Kibuye, Gikongoro, and Cyangugu overall have the greatest soil losses, large soil losses (>5 t/ha/season) were monitored in six of the ten prefectures. Sixteen percent of the sampled fields appear to be having major erosional losses during the first growing season from an agricultural perspective. During the second growing season, both rainfall and soil losses were less. In the second season, only four fields had more than 5 t/ha of soil loss (tab.5). Three were located in Gikongoro and one in Gisenyi.

Tab.6 lists the properties of the twenty field sites where large soil losses were measured. The average slope for these fields is 17 degrees (31%). This is slightly greater than the 13°15" mean slope value of all the sampled fields. Eighteen of the twenty high soil loss fields are cultivated in seasonal crops. The two fields having high erosion and under perennial crops are very steep, 23 and 26 degrees respectively. Finally, a comparison of fig.1 with tab.6 indicates that the areas of new settlement in the eastern portion of the country are not experiencing any large soil losses.

In comparison to soil loss measurements undertaken in Kenya under a similar range of environmental settings and using the identical methodology (LEWIS 1985), the erosion problem appears significantly less in Rwanda. In agreement with the Kenyan findings, the greatest erosion overwhelmingly occurs in fields cultivated in annual crops. Likewise, inspection of the weekly soil loss values indicates that the greatest soil loss normally occurs in the early portion of each growing season; a rain event of a given intensity occurring early in the growing season results in higher soil losses than the same magnitude rain event later in the season. Likely this reflects the poor groundcover and loose soils, resulting from field preparation, that do not dampen the potential erosivity of the rainfall prior to striking the soil. Additionally, a few rain events produce the majority of the soil losses. Contrasting to the Kenyan results, the newly settled areas in Rwanda are not experiencing high soil losses. Also, the steep agricultural lands in Rwanda are clearly experiencing lower erosional rates than found in the central highlands of Kenya.

While insufficient data exist from this study to substanciate the reasons why soil loss values are less in Rwanda, two likely reasons are put forward to explain this behaviour. First, unlike the Kenya fields which are generally cultivated in single crops, almost all fields in Rwanda

Soil Loss, Rwanda

Préfecture	Commune	Slope (degrees)	Crop(s)	Soil Loss (t/ha)
Butare	Muyaga	10	Beans/maize	13.6
Butare	Muyaga	17	Beans/manioc	6.0
Cyangugu	Karengera	14	Beans/maize	12.1
Cyangugu	Kerengera	23	Manioc/cocoyam	9.6
Gikongoro	Kinyamakara	18	Manioc/quinine	13.3
Gikongoro	Rwamiko	10	Eleusine	6.4
Gikongoro	Rwamiko	15	Cocoyam	17.0
Gikongoro (2)	Kinyamakara	18	Manioc/quinine (H)	11.2
Gikongoro (2)	Kinyamakara	20	Sorghum	5.0
Gikongoro (2)	Rwamiko	15	Eleusine	5.7
Gisenyi	Karago	14	Maize/potato	5.6
Gitarama	Mugina	10	Beans/maize	13.7
Gitarama	Nyamabuye	9	Banana/beans	5.1
Kibuye	Kivumu	11	Beans/potato	6.0
Kibuye	Kivumu	23	Bananas	4.9
Kibuye	Kivumu	26	Bananas	6.2
Kibuye	Gitesi	22	Beans/maize	4.8
Kibuye	Gitesi	18	Beans/maize	7.2
Kibuye	Gitesi	20	Manioc/sweet potato	8.4
Ruhengeri	Nkuli	27	Potato	5.3

(2) = second growing season
(H) = harvested during the season

Tab. 6: *Properties of field sites having high erosion losses per growing season.*

are intercropped. Second, as the staple crop of the country is bananas, which provides a good groundcover throughout the year compared to maize in Kenya, a smaller prcentage of bare soil is exposed to the direct impact of the rains in Rwanda. While mechanical conservation practices such as terracing and cut-off drains are found to a far greater degree in Kenya, the protection of the soil from the direct impact of the rain appears to be the single most critical factor limiting soil losses in the highland environments of Kenya and Rwanda. Support is given to this finding through examination of the soil loss values of fields cultivated in coffee. All coffee fields in Rwanda are very heavily mulched. This conservation practice is encouraged by the government to increase coffee yields and hence maximize foreign currency earnings from coffee exports. For the sixteen fields monitored in coffee, all soil losses were less than 1 t/ha per growing season. These low losses occurred throughout the country on fields ranging from 8 to 30 degrees in steepness. Thus the practice of mulching, which has been introduced largely for economic reasons to increase soil moisture for improving coffee yields, is a very successful soil conservation practice too.

4.2 Estimating Soil Loss in Rwanda

The USLE was developed using data collected in the U.S.A. (WISCHMEIER & SMITH 1978). Because of data constraints in general, and significant differences in the agricultural systems, the steepness of land under cultivation, and conservation practices from the phenomena upon which the USLE were developed, it is necessary to assemble the available and field collected data in a format that is valid for applying the equation in the Rwandaise context. The USLE is defined as:

$$A = 2.25\ RKLSCP$$

where

- A = average annual soil loss in mt/ha
- R = rainfall - erosivity factor
- K = soil erodibility factor
- L = slope length factor
- S = slope steepness factor
- C = crop cover and management factor, and
- P = conservation practice factor.

4.3 Estimating Erosivity

Actual measurements of erosivity are exceedingly few for the country as a whole, therefore erosivity for all sample fields was inferred using the rainfall data collected at the nearest rainfall station by the national weather service. Estimated annual R-values were derived using the general equation developed by MOORE (1979) for East Africa and applied successfully in Kenya (LEWIS 1985). An R-value was computed for every field for each growing season. The first growing season's (Nov.–April) R-values ranged from a high of 250 in southern Cyangugu to a low of 150 in eastern Byumba. During the second growing season (May–Oct.) the high value was 135 in northern Gisenyi; the low value was 79 found both in southern Gikongoro and eastern Kigali. In general R-values for the first growing season were double those of the second season. In all cases the erosivity values were larger for the first season at all 100 sample sites. Since the rainfall data used in the study are from local weather stations near the sites, but not from the sample fields themselves, the R-values used in this study are approximations of the actual erosivity experienced at each field.

4.4 Estimating Erodibility

Parent material	K-value
Alluvium	0.20
Basalts	0.12
Colluvium	0.22
Granites	0.20
Volcanics	0.12
Quartzites and schists	0.18
Schists and quartzites	0.15

Tab. 7: *Estimated erodibility values.*

At the time of this study, no published data for representative K-values existed for soils throughout the country. Fortunately almost all soil properties in Rwanda are closely related to the local geology (parent material). Using the soil information developed at the agricultural experimental station in Rabona as well as discussing the problem with individuals having field experience in the country, the following K-values were estimated for the major parent materials found throughout the country (tab.7). The parent material for each site was obtained from direct observation and available geologic information (MINISTERE DES RESSOURCES NATURELLES 1981).

4.5 Slope-Length Factors

The slopes and lengths of each field were measured directly in the field. In the computation of the USLE these two variables are combined into a single topographic factor (LS) where

LS = $(\lambda/72.6)^m (65.41 \sin^2 \theta + 4.56 \sin \theta + 0.065)$ where

λ = slope length and

θ = angle of slope and

m = a constant determined by the slope angle.

Because steep slopes are cultivated in Africa far in excess of what was evaluated for American farming conditions (WISCHMEIER & SMITH 1978, 12), the m exponent values in the LS equation were extended (tab.8). these are the identical values utilized in the Kenyan study.

Slope (Degrees)	M Exponent
0.0– 1.0	0.2
1.1– 1-5	0.3
1.6– 2.0	0.4
2.1– 7.0	0.5
7.1–14.0	0.6
greater than 14.0	0.7

Tab. 8: *M exponent values for topographic (LS) factor.*

4.6 Estimating the Crop Cover and Conservation Practice Factors

A total of 47 diffrent crop combinations were monitored on the 200 sample fields during the one year study period. Some crops such as beans and maize are primarily grown during the first season; other crops such as sorghum are concentrated in the drier second season. The following C-values (crop cover) were assigned for the observed crops based on published C-values for Kenya (LEWIS 1985) and the measured soil loss occurring on the fields having these plant covers (tab.3). These C-values while similar to those for Kenya, do differ. Likewise, these values often differ greatly from those published in the U.S.A. Tobacco's 0.45 value is significantly larger in Rwanda. This reflects the differences in agricultural practices between the heavily subsidized commercial tobacco crops in the U.S.A. and the largely subsistence crops in Rwanda. In Rwanda the soils in tobacco fields are exposed to raindrop impact to a far larger percentage because of lower planting densities.

Tab.3 indicates the relative roles of different crops and crop combinations in protecting the soil from erosion. The soil conserving crops are those with the lower C-values such as coffee and bananas. Both the canopy and root characteristics of the plants affect the C-values. As such the planting patterns that the individual farmers use will contribute to the soil losses on each field. Tab.2 lists the five groundcover categories and the weights added to each C-value to compensate for the different planting strategies and growing conditions on each of the sample fields.

Eight categories of conservation were observed practiced throughout the country (tab.4). But these eight combinations only result in five groupings sensitive to soil loss. Protection of the soil cover (mulching) appears to be the single most successful method of minimizing soil loss. This is fortunate as it is the least capital and labor intensive practice. Likely it is the most cost effective strategy of controlling agricultural erosion in the country.

4.7 Soil Loss Estimate Results

Using the methodology just presented, soil loss values during each growing season were estimated for each sample field.

Correlation coefficients between soil loss estimated by the USLE and measured soil loss were 0.71 (n=99) for the first season and 0.62 (n=97) in the second season. Both correlations are significant at the 99% level. Explained variation (r^2) is 50 and 37 percent respectively. Using a much better data base then available in Rwanda, predictions of the USLE in the U.S.A. only approximate 80 percent. Thus as the first step in developing an estimate of soil loss conditions for Rwanda, this study's methodology is promising.

Some of the unexplained variation in the soil loss estimates must reflect shortcomings in the data base. In particular erosivity values were estimated using precipitation values obtained from the nearest weather stations. Given the local nature of tropical rains, extrapolating these values to specific sites introduces errors in the R-alue input. Additionally the R-values were generated using equations (MOORE 1979); these only approximate erosivity. While improvement in the meteorological data base would improve the prediction of soil loss, precipitation data constraints are likely to remain.

The need to approximate erodibility is an additional source of unexplained variation in the soil loss estimates. With the recent completion of the natural soil mapping project, it will be possible to revise the methodology for assigning K-values. This should improve the estimates of soil loss. Recognizing the limitations of the results, this study's methodology is being utilized to estimate soil loss for the 20.000 fields that comprise the sample population of the complete agricultural survey. The results of this endeavor will provide a better picture of Rwanda's soil loss patterns.

5 Conclusions

The data in this study indicate that three préfectures have relatively severe soil losses. Gikongoro has the greatest overall soil loss problem. The results indicate that methods of cultivation which maintain a crop cover are very successful in minimizing erosion on agricultural lands, even those that are exceedingly steep. From the environmental perspective, the traditional agricultural practices are very well suited for the local landscapes. Fortunately, the traditional farming strategies are continuing to be used in the eastern areas of new agricultural settlement. The result is that soil losses in these lower potential areas are minimal. This contrasts to Kenya where the soil resource is being 'mined' in most newly settled areas even when terracing and cut-off trenches exist. Encouraging intercropping, perennial crops, and application of mulches are very effective management policies for minimizing soil losses for the conditions found throughout Rwanda. It is likely they would be equally successful in other tropical highland countries.

To obtain an overview of the national and regional soil loss patterns, the USLE provides a useful methodology. However, the topographic component of the equation had to be revised. First, it was necessary to modify the slope factor because of the steeper lands under cultivation. And second, slope length only affected soil losses up to field lengths of 15 meters. The assumption that runoff and hence soil loss increases with field length appears invalid for distances in excess of 15 meters in Rwanda. Comparison of the estimated values to the measured soil loss data was significant. While large differences exist for some specific fields between the estimated and measured values, the overall pattern of soil loss at the préfecture and commune scales is duplicated by the modified USLE. For planning purposes, the methods used in this study for asigning erosivity and erodibility values are valid. But they should not be used to establish site specific recommendations for individual fields.

The assignment of specific C-values for the major crop covers found in Rwanda indicate the extreme caution that must be used when applying the USLE in tropical areas. In most cases the locally derived values differ greatly from the widely published values determined under different agricultural systems. Inspection of the derived conservation factor values (P-values) indicate that engineering methods of erosion control are not as effective as mulching. This result could indicate that either terraces or cut-off trenches have been poorly designed for local conditions or that they are not properly maintained. Clearly for the Rwandaise situation, appropriate conservation strategies should initially minimize engineering solutions. For the conditions found throughout the country, this study's results indicate that the maintenance of a good groundcover is the single most important factor in controlling erosion.

Acknowledgement

The research reported in this paper was funded by L'Enquête Nationale sur l'Agriculture au Rwanda, Ministère de l'Agriculture, de l'Elevage et des Fôrets and the United States Agency for International Development.

References

BUREAU DE RECENSEMENT (1982): Recensement Général de la Population et de l'Habitat, 1978, 57 p.

DIVISION DE CLIMATOLOGIE (1984): Bulletin Climatologique Annee 1983. 88 p.

DIVISION DE CLIMATOLOGIE (1985): Bulletin Climatologique Annee 1984. 81 p.

FELDMAN, W.M., FREEMAN, P., ROURKE, J.O. & McGAHUEY, M. (1985): Evaluation of the Environmental Training and Management Project. Office of Regional Affairs, Bureau for Africa, A.I.D., Washington, D.C.

LEWIS, L.A. (1985): Assessing Soil Loss in Kiambu and Murang'a Districts, Kenya. Geografiska Annaler, **67 A**, 273–284.

LEWIS, L.A. & COFFEY, W.J. (1985): The Continuing Deforestation of Haiti. Ambio, **XIV**, 158–160.

MINISTERE DES RESSOURCES NATURELLES (1981): Carte Lithologique Du Rwanda.

MINISTERE DE L'AGRICULTURE ET DE L'ELEVAGE (1983a): Quelques Resultats de la Phase Pilote. 72 p.

MINISTERE DE L'AGRICULTURE ET DE L'ELEVAGE (1983b): Resultats Preliminaires de la Phase Pilote. 18 p.

MOORE, T.R. (1979): Rainfall erosivity in East Africa. Geografiska Annaler, **61 A**, 147–156.

REPETTO, R. (1986): Soil Loss and Population Pressure on Java. Ambio, **XV**, 14–18.

WASSMER, P. (1981): Etude de l'érosion des sols et de ses conséquences dans la Prefecture de Kibuye. Thèse de doctorat, Universite Louis Pasteur, Strasbourg. 158 p.

WISCHMEIER, W.H. & SMITH, D.D. (1978): Predicting Rainfall Erosion Losses — A Guide to Conservation Planning. U.S. Department of Agriculture, Washington. 58 p.

WORLD BANK (1983): Rwanda Economic Memorandum — Recent Economic and Sectoral Developments and Current Policy Issues. Report No. **4059-RW**, 36 p.

Address of author:
L.A. Lewis
Graduate School of Geography
Clark University
Worcester, Massachusetts 01610
U.S.A.

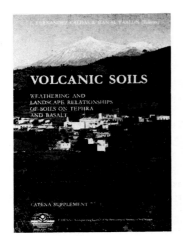

E. Fernandez Caldas & Dan H. Yaalon (Editors):

VOLCANIC SOILS
Weathering and Landscape
Relationships of Soils on Tephra and Basalt

CATENA SUPPLEMENT 7, 1985

Price DM 128,—

ISSN 0722–0723 / ISBN 3-923381-06-9

PREFACE

This CATENA SUPPLEMENT contains selected papers presented at the International Meeting on Volcanic Soils held in Tenerife, July 1984. The meeting brought together over 80 scientists from 21 countries, with interest in the origin, nature and properties of soils on tephra and basaltic parent materials and their management. Some 51 invited and contributed papers and 8 posters were presented on a wide range of subjects related to volcanic soils, many of them dealing with weathering and landscape relationships. Classification was also discussed extensively during a six day excursion of the islands of La Palma, Gomera and Lanzarote, which enabled the participants to see the most representative volcanic soils of the Canary Archipelago under a considerable range of climatic regimes and parent material ages.

Because volcanic soils are not a common occurrence in regions where pedology developed and progressed during its early stages, recognition of their specific properties made an impact only in the late forties. The name **Ando** soils, now recognized as a special Great Group in all comprehensive soil classification systems, was coined in 1947 during reconnaissance soil surveys in Japan made by American soil scientists. Subsequently a Meeting on the Classification and Correlation of Soils from Volcanic Ash, sponsored by FAO and UNESCO, was held in Tokyo, Japan, in 1964, in preparation for the Soil Map of the World. This was followed by meetings of a Panel on Volcanic Ash Soils in Latin America, Turrialba, Costa Rica, in 1969 and a second meeting in Pasto, Colombia, in 1972. At the International Conference on Soils with Variable Charge, Palmerston, New Zealand, 1981, the subject of Andosols was discussed intensively. Most recently the definitions of Andepts, as presented in the 1975 U.S. Soil Taxonomy, prompted the establishment of an International Committee on the Classification of Andisols (ICOMAND), chaired by M. Leamy from C.S.I.R., New Zealand, which held a number of international classification workshops, the latest in Chile and Ecuador, in January 1984. The continuous efforts to improve and revise the new classification of these soils is also reflected in some of the papers in this volume.

While Andosols or Andisols formed on tephra (volcanic ash), essentially characterized by low bulk density (less than 0.9 g/cm^3) and a surface complex dominated by active Al, cover worldwide an area of about 100 million hectares (0.8% of the total land area), the vast basaltic plateaus and their associated soils cover worldwide an even greater area, frequently with complex age and landscape relationships. While these soils do not generally belong to the ando group, their pedogenetic pathways are also strongly influenced by the nature and physical properties of the basalt rock. The papers in this volume cannot cover the wide variety of properties of the soils in all these areas, some of which have been reviewed at previous meetings. In this volume there is a certain emphasis on some of the less frequently studied environments and on methods of study and characterization as a means to advance the recognition and classification of these soils.

The Tenerife meeting was sponsored by a number of national and international organizations, including the Autonomous Government of the Canary Islands, the Institute of Ibero American Cooperation in Madrid, the Directorate on Scientific Policy of the Ministry of Education and Science, Madrid, the International Soil Science Society, ORSTOM of France, and ICOMAND. Members and staff of the Department of Soil Science of the University of La Laguna had the actual task of organizing the meeting and the field trips. In editing the book we benefitted from the manuscript reviews by many of our colleagues all over the world, and the capable handling and sponsorship of the CATENA VERLAG. To all those who have extended their help we wish to express warm thanks.

La Laguna and Jerusalem, E. Fernandez Caldas
Summer 1984 D.H. Yaalon
  Editors

CONTENTS

R.L. PARFITT & A.D. WILSON
ESTIMATION OF ALLOPHANE AND HALLOYSITE IN THREE SEQUENCES OF VOLCANIC SOILS, NEW ZEALAND

J.M. HERNANDEZ MORENO, V. CUBAS GARCIA, A. GONZALEZ BATISTA & E. FERNANDEZ CALDAS
STUDY OF AMMONIUM OXALATE REACTIVITY AT pH 6.3 (Ro) IN DIFFERENT TYPES OF SOILS WITH VARIABLE CHARGE. I

E. FERNANDEZ CALDAS, J. HERNANDEZ MORENO, M.L. TEJEDOR SALGUERO, A. GONZALEZ BATISTA & V. CUBAS GARCIA
BEHAVIOUR OF OXALATE REACTIVITY (Ro) IN DIFFERENT TYPES OF ANDISOLS. II

D.J. RADCLIFFE & G.P. GILLMAN
SURFACE CHARGE CHARACTERISTICS OF VOLCANIC ASH SOILS FROM THE SOUTHERN HIGHLANDS OF PAPUA NEW GUINEA

J. GONZALEZ BONMATI, M.P. VERA GOMEZ & J.E. GARCIA HERNANDEZ
KINETIC STUDY OF THE EXPERIMENTAL WEATHERING OF AUGITE AT DIFFERENT TEMPERATURES

P.A. RIEZEBOS
HIGH-CONCENTRATION LEVELS OF HEAVY MINERALS IN TWO VOLCANIC SOILS FROM COLOMBIA: A POSSIBLE PALEOENVIRONMENTAL INTERPRETATION

L.J. EVANS & W. CHESWORTH
THE WEATHERING OF BASALT IN AN ARCTIC ENVIRONMENT

R. JAHN, Th. GUDMUNDSSON & K. STAHR
CARBONATISATION AS A SOIL FORMING PROCESS ON SOILS FROM BASIC PYROCLASTIC FALL DEPOSITS ON THE ISLAND OF LANZAROTE, SPAIN

P. QUANTIN
CHARACTERISTICS OF THE VANUATU ANDOSOLS

P. QUANTIN, B. DABIN, A. BOULEAU, L. LULLI & D. BIDINI
CHARACTERISTICS AND GENESIS OF TWO ANDOSOLS IN CENTRAL ITALY

A. LIMBIRD
GENESIS OF SOILS AFFECTED BY DISCRETE VOLCANIC ASH INCLUSIONS, ALBERTA, CANADA

M.L. TEJEDOR SALGUERO, C. JIMENEZ MENDOZA, A. RODRIGUEZ RODRIGUEZ & E. FERNANDEZ CALDAS
POLYGENESIS ON DEEPLY WEATHERED PLIOCENE BASALT, GOMERA (CANARY ISLANDS): FROM FERRALLITIZATION TO SALINIZATION

SOIL LOSS AND SEASONAL VARIATION OF ERODIBILITY IN TWO SOILS WITH DIFFERENT TEXTURE IN THE MUGELLO VALLEY IN CENTRAL ITALY

C. **Zanchi**, Firenze

Summary

The results of monthly soil loss measurements for 1978–1983, obtained on 4 plots located on soils of different texture, are reported.

The erodibility value for each month was evaluated and it shows that these vary in the different months according to a cosine function.

It was also observed that the erodibility values, even though highly varying for the two different types of soil, had a similar pattern.

The values of the corrective factors, for both soils, obtained by dividing the erodibility of each month by the average yearly value, did not show, statistically significative differences.

The obtained results indicate the possibility of extending the U.S.L.E. equation to the forecast of average soil loss on a monthly base.

ISSN 0722-0723
ISBN 3-923381-12-3
©1988 by CATENA VERLAG,
D–3302 Cremlingen-Destedt, W. Germany
3-923381-12-3/88/5011851/US$ 2.00 + 0.25

Resumen

Uno de los métodos más conocidos y extendidos para predecir la pérdida do suelo es la Ecuación Universal de la la Pérdida de Suelo (USLE), la cual permite estimar la pérdida media de suelo en un período largo de tiempo. Pero la ecuación no es muy adecuada para predecir pérdidas de suelo en períodos más cortos de un año. La erodibilidad (K) depende del suelo y en la Ecuación se considera como una constante, de manera que las variaciones en las condiciones de humedad y en algunas características físicas del suelo a lo largo del año no son tomadas en consideración.

El conocimiento de la variación de la erodibilidad a lo largo del año permitiría usar la Ecuación para predecir pérdidas de suelo a escala mensual. Además se podría mejorar la precisión del método mediante la asociación de varios valores de erodibilidad con los estadios de la cobertura vegetal de las plantaciones y del factor de cultivo (C).

El objetivo de nuestra investigación es el de estudiar las variaciones de la erodibilidad del suelo durante varios meses, y verificar si tales variaciones son similares en suelos con diferentes texturas pero localizados entre si a una distancia inferior a los 200 metros, y por tanto con el mismo microclima.

1 Introduction

One of the best known and most widespread methods of forecasting soil loss is with the "Universal Soil Loss Equation" (WISCHMEIER & SMITH 1978).

This statistical model is:

$$A = R.K.LS.C.P. \qquad (1)$$

where A is the soil loss expressed in t/ha, R is the rain erosivity in MJmm/ha.h.y, K is the soil erodibility in t.ha.h/MJha.mm, L and S are the dimensionless slope length and steepness factor, C and P are respectively the dimensionless cover and management and support practice factors; it allows the average soil loss for a long period of time to be forecast. It is not so reliable when used to forecast soil losses, for periods, shorter than one year (ZANCHI 1978). Soil erodibility (K) is a long term average and so considered constant throughout the year so that variations in moisture conditions and of some physical characteristics of the soil during the year, are not taken into consideration.

If it is known how soil erodibility varies during the year, the U.S.L.E. method could be used to forecast monthly soil losses.

Furthermore it would improve the precision of this method by associating the various erodibility values with the respective cropstages of the cover and management factor (C).

The goal of this research is to study the variations of soil erodibility during the various months, and at least during this first stage, to verify whether such variations are similar on two soils with different texture, located at two places less than two hundred metres apart having the same micro-climate.

2 Experimental Methodology

The study was carried out on four bare fallow plots located on two neighbouring slopes, characterized by soils with different textures, and with 10% slope.

The average yearly rainfall of this area is 1051 mm, having its minimum in July and two maximum peaks in November and February; the rainfall was measured by a recording raingauge located near the plots.

The erosivity of the rain was calculated according to WISCHMEIER & SMITH (1978).

The average rainfall erosivity of the area is 2038 MJmm/ha.h.y. To calculate the erodibility, we have taken into consideration the erosivity values related to the rain events, with more than 7.0 mm, which is the experimental threshold of the erosive rains in such area (ZANCHI 1978). At the same time we have not taken into consideration those events for which we did not have rainfall or soil loss data. The average yearly temperature was 13.4°C; the minimum temperature being in January and the maximum in July/August.

Two of the plots have soils classified as "aeric fluvaquents", and the other two "typical eutrocrepts" (tab.1).

The plots, (5 × 20 metres), arranged with the major axis in the direction of the maximum inclination, were bounded by bourders set 15 cm into the ground and protruding 20 cm above the ground.

After each rainfall event the quantity

Type of soil	clay	silt	sand	O.M.	pH(H$_2$O)	CaCo$_3$
FLUVAQUENTS	21.7	40.6	37.7	1.8	7.8	n.d.
EUTROCREPTS	54.2	40.7	5.1	1.6	8.2	10.5

Tab. 1: *Som physical and chemical characteristics of the two soils.*

year	Runoff Eutrocrepts mm	Runoff Fluvaquents mm
1978	97.59	142.01
1979	392.53	216.08
1980	302.52[4]	184.90
1981	508.91[4]	142.92[3]
1982	614.65[5]	44.80[1]
1983	215.51	178.61[2]
Total	2,131.71	909.32

[1] Missing data for the months from August to December.
[2] Until August.
[3] Missing data for September.
[4] Missing data for July, August and September.

Tab. 2: *The yearly average runoff of the two soils.*

of runoff was measured and sampled to establish the concentration of eroded soil.

The plots were kept bare fallow by ploughing with a major Summer up and down slope tillage, followed by secondary work to place the plots in conventional seedbed conditions.

3 Results and Discussion

The results presented here, were collected between December 29th, 1977 and August 3rd, 1983.

The annual average runoff of the two replications (tab.2), demonstrate a large difference between the two types of soil.

In order to study the variability of the erodibility during the year the soil losses of each rainfall were considered monthly.

The average soil loss of the two replications (A) during the various months and the related erodibility values are shown in tab.3 and 4 respectively.

Because of the different length and slope of the experimental plots, the erosion values of tab.3 and 4 have been related, through the LS factor, to those of the standard plots (22.13 m long; 9% steep).

The erodibility values calculated for each month are the average value of that specific time of the year. In fact the erodibility standard deviations are large and, in some cases, larger than the monthly averages, demonstrating a wide variation, in each month, of the erosion and for the erosivity.

In tab.3 and 4 the values of the corrective factor are also presented, to enable the evaluation of the erodibility for each month starting from the average yearly erodibility value (K). The corrective factor has been obtained dividing the value of the monthly erodibility by the value of the yearly one. The corrective factor could be applied to the K factor of the U.S.L.E. to forecast the soil loss on a monthly base.

The erodibility values, even though they substantially differ in the two types of soil, show a similar pattern in the course of the various months.

On both soils, the erodibility value is

Months	A t/ha	average Km t.h/MJ.mm	Std.Dev.	Kc** (corrective factor)
January	28.803	0.041	0.048	1.519
February	22.903	0.049	0.069	1.815
March	27.540	0.040	0.028	1.481
April	5.018	0.008	0.015	0.296
May	1.414	0.006	0.009	0.222
June	16.485	0.012	0.011	0.444
July[1]	1.340	0.009	0.007	0.333
August[2]	2.836	0.003	0.001	0.111
September[1,2]	0.718	0.002	0.001	0.074
October	30.160	0.054	0.045	2.000
November	39.171	0.033	0.033	1.222
December	56.304	0.042	0.057	1.566
Annual (Ka)		0.027		

[1] Missing data for 1981
[2] Missing data for 1982
** Kc comes from Km/Ka

Tab. 3: *Soil loss and monthly erodibility (Km) for the eutrocrepts soil on location "Crocioni"*.

Months	A t/ha	average Km th/MJmm	Std.Dev.	Kc** (corrective factor)
January	4.270	0.0075	0.0052	1.4286
February	1.825	0.0058	0.0022	1.1048
March[1]	6.439	0.0081	0.0068	1.5429
April	2.288	0.0038	0.0033	0.7238
May	0.064	0.0007	0.0006	0.1333
June	2.145	0.0017	0.0012	0.3238
July[1]	0.584	0.0024	0.0040	0.4571
August[1,2]	0.763	0.0007	0.0005	0.1333
September[1,2]	0.988	0.0024	0.0026	0.4571
October[2]	2.608	0.0080	0.0044	1.5238
November[2]	15.047	0.0136	0.0186	2.5905
December[2]	8.036	0.0083	0.0067	1.5810
Annual (Ka)		0.0053		

[1] Missing data for 1981
[2] Missing data for 1982
** Kc comes from Km/Ka

Tab. 4: *Soil oss and monthly erodibility (Km) for the fluvaquents soil on location "Stalla"*.

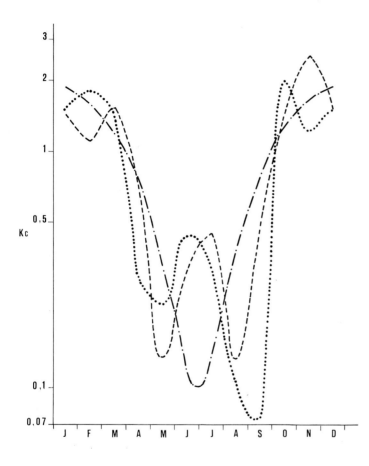

Fig. 1: *Trend of corrective factor (K_c) for locations Stalla, Crocioni and calculated with equation (4).*

at its peak during the months between October and March, while it decreases to much lower values during the rest of the year.

The maximum erodibility value, in the months of October and November, is certainly influenced by the preparation of the seed-bed that makes the soil particularly susceptible to erosion.

In contrast, during the Summer, even after very strong rainstorms soil loss is generally limited, mainly because of the many shrinkage cracks that are typical of these soils.

The statistical analysis of the corrective factors has in fact shown that the difference of the values found in the two places is not significant.

The pattern of the values of the corrective factors (Kc) during the months of the year, considered respectively as dependent and independent variables, has been studied through polynomial regression.

The results of such a regression have shown that the best fit has been obtained with the following equations characterized by high coefficients of determination (R^2).

"Stalla" area:

$$Kc = 2.3818 - 1.6330x + 0.8789x^2 - 0.2177x^3 + 0.0225x^4 - 0.0008x^5$$
$$R^2 = 0.86 \qquad (2)$$

"Crocioni" area:

$$Kc = 0.5480 + 1.6910x - 0.7562x^2 + 0.1089x^3 - 0.0060x^4 + 0.00009x^5$$
$$R^2 = 0.71 \qquad (3)$$

The corrective factors have been then interpolated also with a function of the type:

$$Kc = 1 + 0.90\cos(t.30°) \qquad (4)$$

where t is the number of each month (0.5 = January 15th; 1.5 = February 15th etc.) . Such a function has been reported previously (ZANCHI 1983). The cosinusoidal trende takes into consideration the cyclic pattern of erodibility during the year.

From fig.1, it can be seen that there is a relationship, particularly during the Winter, between the values of the corrective factor (Kc), found for locations Stalla and Crocioni, and those calculated with equation (4). The deviation in Summer months between measured and calculated curves is mainly due to the low value of soil erodibility caused by the large and diffuse cracks.

4 Conclusion

The experimental results demonstrate the pronounced variability of erodibility throughout the year. At Stalla we have noticed, compared to the average yearly value, a fluctuation between 13% in August and 256% in November.

At Crocioni the data show a fluctuation, compared to the average value, between an minimum of 7% in September and a maximum of 200% in October. Such a variation is modelled, with a cosine function with which it is possible to predict fluctuations.

It seems important that in both soils a similar pattern of cosinusoidal fluctuation was observed, even though the soils had very different erodibility values. This was confirmed by the non-significative difference of the monthly corrective factors (Kc) that were found for the plots under study.

The cosinusoidal trend of the corrective factors is similar to that found in North Mississippi (U.S.A.) by MUTCHLER & CARTER (1983) for other different types of soil. The use of the corrective factors seems so of particular importance to extende the U.S.L.E. validity to forecast the average monthly soil loss.

Certainly this type of trend must be verified also for other types of soil, but these first results seem promising.

References

MUTCHLER, C.K. & CARTER, C.E. (1983): Soil erodibility variation during the year. Transaction of the ASAE, 1102–1104, 1108.

WISCHMEIER, W.H. & SMITH, D.D. (1978): Predicting rainfall erosion losses — A guide to conservation planning. U.S.D.A., Agricultural Handbook n. **537**, 1–58.

ZANCHI, C. (1978): Predicting the soil losses and the soil concentration in the runoff in function of some rain physical characteristics and of a runoff factor. (In Italian) Annali istituto Sperimentale per lo studio e la difesa del suolo, vol. **IX**, 217–230.

ZANCHI, C. (1983): Relative influence of rain splash and runoff on soil loss, and soil erodibility variation during the year. (In Italian) Annali Istituto Sperimentale per lo Studio e la Difesa del Suolo, Vol. **XIV**, 347–358.

RESEARCH WORK SUPPORTED BY C.N.R.
ITALY.
SPECIAL GRANT I.P.R.A.
SUB-PROJECT 2 — PAPER no. 1043.

Address of author:
C. Zanchi
Instituto do Agronomia Generale e
Coltivazioni erbacee
Università di Firenze
Firenze, Italy

DAN H. YAALON (ED.)

ARIDIC SOILS and GEOMORPHIC PROCESSES

SELECTED PAPERS of the INTERNATIONAL CONFERENCE
of the INTERNATIONAL SOCIETY of SOIL SCIENCE
Jerusalem, Israel, March 29 – April 4, 1981

CATENA SUPPLEMENT 1, 1982

Price: DM 95,–

ISSN 0722–0723 / ISBN 3–923381–00–X

This CATENA SUPPLEMENT comprises 12 selected papers presented at the International Conference on Aridic Soils – Properties, Genesis and Management – held at Kiryat Anavim near Jerusalem, March 29 – April 4, 1981. The conference was sponsored by the Israel Society of Soil Science within the framework of activities of the International Society of Soil Science. Abstracts of papers and posters, and a tour guidebook which provides a review of the arid landscapes in Israel and a detailed record of its soil characteristics and properties (DAN et al. 1981) were published. Some 49 invited and contributed papers and 23 posters covering a wide range of subjects were presented at the conference sessions, followed by seven days of field excursions.

The present collection of 12 papers ranges from introductory general reviews to a number of detailed, process oriented, regional and local studies, related to the distribution of aridic soils and duricrusts in landscapes of three continents. It is followed by three papers on modelling and laboratory studies of geomorphic processes significant in aridic landscapes. It is rounded up by a methodological study of landform–vegetation relationships and a regional study of desertification. Additional papers, related to soil genesis in aridic regions, are being published in a special issue of the journal GEODERMA.

D.H. Yaalon
Editor

G.G.C. CLARIDGE & I.B. CAMPBELL
   A COMPARISON BETWEEN HOT AND COLD DESERT SOILS AND SOIL PROCESSES

R.L. GUTHRIE
   DISTRIBUTION OF GREAT GROUPS OF ARIDISOLS IN THE UNITED STATES

M.A. SUMMERFIELD
   DISTRIBUTION, NATURE AND PROBABLE GENESIS OF SILCRETE IN ARID AND SEMI-ARID SOUTHERN AFRICA

W.D. BLÜMEL
   CALCRETES IN NAMIBIA AND SE-SPAIN RELATIONS TO SUBSTRATUM, SOIL FORMATION AND GEOMORPHIC FACTORS

E.G. HALLSWORTH, J.A. BEATTIE & W.E. DARLEY
   FORMATION OF SOILS IN AN ARIDIC ENVIRONMENT WESTERN NEW SOUTH WALES, AUSTRALIA

J. DAN & D.H. YAALON
   AUTOMORPHIC SALINE SOILS IN ISRAEL

R. ZAIDENBERG, J. DAN & H. KOYUMDJISKY
   THE INFLUENCE OF PARENT MATERIAL, RELIEF AND EXPOSURE ON SOIL FORMATION IN THE ARID REGION OF EASTERN SAMARIA

J. SAVAT
   COMMON AND UNCOMMON SELECTIVITY IN THE PROCESS OF FLUID TRANSPORTATION:
   FIELD OBSERVATIONS AND LABORATORY EXPERIMENTS ON BARE SURFACES

M. LOGIE
   INFLUENCE OF ROUGHNESS ELEMENTS AND SOIL MOISTURE ON THE RESISTANCE OF SAND TO WIND EROSION

M.I. WHITNEY & J.F. SPLETTSTOESSER
   VENTIFACTS AND THEIR FORMATION: DARWIN MOUNTAINS, ANTARCTICA

M.B. SATTERWHITE & J. EHLEN
   LANDFORM–VEGETATION RELATIONSHIPS IN THE NORTHERN CHIHUAHUAN DESERT

H.K. BARTH
   ACCELERATED EROSION OF FOSSIL DUNES IN THE GOURMA REGION (MALI) AS A MANIFESTATION OF DESERTIFICATION

SOME RESULTS OF SOIL EROSION MONITORING AT A LARGE-SCALE FARMING EXPERIMENTAL STATION IN HUNGARY

L. **Góczán** & A. **Kertész**, Budapest

Summary

For the investigation of soil erosion processes a measurement programme was started in 1982 by the Geographical Research Institute of the Hungarian Academy of Sciences serving both purely scientific and practical purposes. To realize the objectives of the programme three sample areas have been selected. The paper presents results obtained in the first sample area.

The station consisting of six plots is situated on the arable land of a cooperative so that the effect of large-scale farming on soil erosion could be studied. It was established in 1982 NW of Pilismarót (North Hungary) on a strongly eroded slope (sol brun lessivé) exposed to the NNW with an average slope angle of 12–13°. Measurements were usually carried out between April and October but in most cases the measuring equipment had to be removed even within this period to enable the cultivation of the field.

As the measurement period was very dry the effect of high intensity rains could not be observed. However, the results of the measurements prove that low intensity rains also cause considerable runoff and erosion under the above-mentioned circumstances. The runoff measurements are compared with results of rainfall simulation experiments. There is a big difference in runoff values determined by the two types of experiments types, namely to the protecting effect of wheat.

A system of small plots along the slope made it possible to measure redeposition along the slope. The role of the factors controlling erosion is relfected in the results of redeposition measurements.

A chemical analysis of runoff samples points to the loss of fertilizers. Due to lack of top soil on the slope the proportion of Ca^{++} in runoff is relatively high suggesting the erosion of loess and of the accumulated $CaCo_3$ from the C horizon.

Although only 4 years of measurements are available, the results seem to be characterized for the region.

Resumen

Se presentan los resultados de erosión del suelo obtenidos en una estación experimental establecida en 1982, y que consiste en seis parcelas situadas en tierras de cultivo. La vertiente tiene una pendiente de 12° a 13°, está expuesta al

ISSN 0722-0723
ISBN 3-923381-12-3
©1988 by CATENA VERLAG
D–3302 Cremlingen-Destedt, W. Germany
3-923381-12-3/88/5011851/US$ 2.00 + 0.25

NNW y el suelo es del tipo pardo lixiviado. Las mediciones se han llevado a cabo de abril a octubre, aunque en muchos casos las instalaciones han debido ser removidas a causa de las faenas de cultivo.

Puesto que el período de mediciones coincidió con un período climático muy seco, no han podido investigarse los efectos de las lluvias intensas. No obstante, los resultados muestran que las lluvias de poca intensidad causan también en estas parcelas una notable escorrentía y erosión. La comparación de estas mediciones de escorrentía con los resultados obtenidos mediante experimentos con simulación de lluvias muestran la existencia de grandes variaciones entre los valores de escorrentía producidos bajo diferentes cultivos.

Un sistema de pequeñas parcelas a lo largo de la vertiente ha servido para medir la redeposición de sedimentos, con lo cual se ha podido ver el papel de los factores que controlan la erosión. El análisi químico de muestras del agua de escorrentía indica un lavado de fertilizantes; la proporción de Ca^{++} es relativamente alta debido a la falta de suelo en la vertiente a causa del lavado de loess y de la acumulación de $CaCO_3$ en el horizonte C. Aunque sólo se poseen 4 años de mediciones los resultados parecen ser una muestra característica de la región.

1 Introduction

The quantitative assessment of soil erosion is very important in Hungary where the soil is one of the most important natural resources and hilly regions with eroded soils are also under cultivation. According to STEFANOVITS (1964) 25% of the total area of Hungary is affected by soil erosion. In order to elaborate effective soil conservation techniques soil erosion forms and processes and the factors controlling erosion have to be studied. For their investigation a measurement programme was started in 1982 by the Geographical Research Institute of the Hungarian Academy of Sciences serving both scientific and practical purposes (KERTESZ 1983). The scientific objectives of the programme include both measuring of **soil erosion processes** (soil loss and the amount of washeddown aggregates sorted by diameter, processes of redeposition along the slope, the amount of surface runoff, fertilizer content of the eroded soil and of the suspension) **under controlled conditions** and **the investigation of the role of the factors influencing soil erosion** (slope angle and exposition, rainfall intensity, soil and rock type, type of cultivation and crop). the practical aspect is manifested in realizing a more successfull planning of the reduction of surface runoff and prevention of soil and fertilizer losses as well as the minimazition of environmental pollution due to sedimentation.

To realize the objectives of the programme three sample areas have been selected. The paper presents the results obtained in the first sample area.

After ERÖDI et al. (1965) a soil loss of 50 million m^3 per year can be estimated for Hungary, which corresponds to an average erosion of 1 mm/year. As soil erosion processes are controlled by several factors soil erosion rates vary not only regionally but also within a small watershed and even along a slope. Estimations over a region or over the whole area of Hungary give only a very rough idea on soil erosion rates so that the importance of measurements and that of the extrapolation of results must be emphasized.

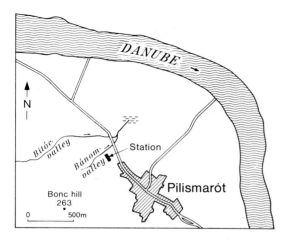

Fig. 1: *Location of the station at Pilismarót.*

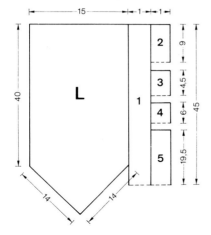

Fig. 2: *The system of plots at the station.*
L = large plot;
1 = control plot;
2-5 = small plots
Lenght and width of the plots are given in meter

2 Location of the Station

The lowest middle section of a slope exposed to the NNW was chosen in the Bánom-valley (fig.1). The intermittent stream discharges into the Danube, allowing conclusions to be made concerning contributions made to the sediment load and to the agrochemical pollution of the Danube.

The station, consisting of 6 plots is situated on the arable land of a cooperative so that the investigation of the effect of large-scale farming on soil erosion should be possible. It was established in 1982 NW of Pilismarót, near the Danube (fig.1). One large and five small plots have been set out (fig.2). The average slope angle along the plot is 12–13°, with a minimum value of 8° on the uppermost section and with a maximum of 15° on the lower part. The surface is covered by an eroded sol brun lessivé typical for the whole area. The upper slope section is, however, very strongly eroded (only the C horizon is available) and the lower section is overlaid by colluvium.

The large plot is 40 m long and 15 m wide (fig.2) ending in a triangle at the lowest section facilitating the collection of water and sediment. Next to the large plot there is a one m wide control plot (photo 2) the length of which corresponds to that of the large plot. On the other side of the large plot 4 small plots (No. 2–5) allow the collection of surface runoff and sediment from each slope section separately (photo 2). The small plots have been installed for the investigation of erosion and deposition processes along the slope. After a thorough field survey the small plots shwon on fig.2 were defined. Lower and upper boundaries of each plot were marked out where the microtopography of the slope changes.

3 Experimentation

The equipment installed on the large and on the small plots is different. The measuring equipment on the large plot was designed by GOCZAN et al. (1973). An inflow pipe connects the plot with the first tank of the equipment consisting of three units (photo 1). Three fractionating sieves with mesh-widths of 2, 0.25 and 0.05 mm have been fixed in the first tank to separate the washed-down aggregates of different diameters. Sieves can be removed to weight the sediment. The sediment remains in the first tank (sludge-tank) the water flows through a dividor to the second and third tanks.

On the small plots, 1 m wide sediment traps after SCHMIDT (1979) have been installed (photo 3). Runoff and sediment are collected in a 20 l subterranean tank placed in front of the trap. the plots are bounded by planks. When the field had to be cultivated the plands and the whole system of plots had to be removed. After cultivation, the planks were relaced. Thus measurements were not carried out continuously.

Tab.1 indicates the management of the field between 1981 and 1985 on which the station is situated.

4 Amount and Frequency of Precipitation

The mean and annual precipitation in the area is 620 mm (50 years average). Tab.2 shows precipitation values between 1978 and 1985. As the measurements were performed between 1982 and 1985 tab.2 must be analysed in detail. Precip-

Photo 1: *Measurement equipment installed in front of the large plot.*

Photo 2: *The system of small plots.*

Photo 3: *Sediment traps constructed after SCHMIDT (1979).*

Time	crop	soil cultivation	manure	fertilizer
1981 Autumn	sunflower late wheat	deep ploughing disk harrowing	60 t/ha	1 t/ha combined NKP fertilizer
1982 Spring	late wheat			0.2 t/ha NH_4NO_3
1982 Autumn	late wheat	ploughing, disk harrowing, frame harrowing		0.8 t/ha combined NPK fertilizer
1983 Spring	late wheat	after harvest deep ploughing (twice)		0.3 t/ha N fertilizer (34% N)
1984 Spring	maize	deep ploughing, harrowing, frame harrowing		0.9 t/ha combined NPK fertilizer
1984 Autumn	maize	deep ploughing		
1985	mustard late wheat	mulching, disk harrowing, ploughing		0.15 t/ha N fertilizer, 0.11 t/ha P and K fertilizer (mixed)

Tab. 1: *Land management on and around the station.*

Year	I	II	III	IV	V	VI	VII	VIII	IX	X	XI	XII	
1978	36.6	28.8	37.6	47.1	68.2	79.8	59.6	31.9	40.3	34.9	17.2	32.7	514.7
1979	81.9	39.8	46.4	54.8	6.5	96.5	43.1	34.5	22.9	14.5	53.6	100.4	594.9
1980	45.2	30.7	35.2	52.2	32.7	69.5	88.1	34.8	27.1	75.6	151.4	34.2	676.6
1981	40.1	11.9	38.5	18.8	74.5	109.6	38.7	74.8	58.1	13.8	52.8	94.5	626.1
1982	65.9	7.3	28.1	4.6	39.3	58.4	28.1	28.1	7.9	42.5	37.5	76.1	424.4
1983	34.8	41.1	18.2	23.3	56.8	60.9	27.0	22.2	30.4	43.4	42.7	13.8	416.6
1984	46.4	19.9	15.2	34.1	83.9	26.8	7.2	5.0	76.4	51.1	55.0	22.1	443.1
1985	10.7	29.9	38.7	20.9	104.2	30.7	15.1	69.5	31.0	13.5	106.2	0.0	470.4
average 1978–1985	50.1	25.6	31.1	33.6	51.7	74.5	46.0	33.0	37.6	39.4	58.6	53.4	528.1

Tab. 2: *Precipitation at the station (1978–1985, mm)*.

itation amounts shown in tab.2 are far below the average values of 1901–1950. 1982 and 1983 were especially dry with an absolute deficit of 200 mm compared to the 50 years average, i.e. two thirds of the 50 years average. As far as the years with relatively higher precipitation are concerned extraordinary high summer values can be observed. The December values couldn't be taken into account as measurements were done only in the summer.

The objective of the measurement programme was to measure after each rainfall event. As some rainy periods lasted 4–5 days it was not always possible to register the erosive effect of each rainfall and the measured runoff and sediment rates were due to a long rainy period with changing intensity (e.g. September 1984).

5 Results

5.1 Runoff

Tab.3 summarizes the results of the measurements between 1982 and 1985. Before analysing the measurements the following facts should be considered.

1. Results of only 4 years measurements are available.

2. Measurements were usually carried out between April and October each year. In most cases the measuring equipment had often to be removed within this period. E.g. 1983 was extremely dry and neither runoff nor sediment could be observed; all the same it must be noted that the measurements begun only in July because of intensive agricultural activity before.

3. There are no measurements during snow melt as the measuring equipment was not used in winter.

4. At the beginning of the measuring period (i.e. 1982 and early 1983) the installation of the small plots was not perfect. That is why in 1982 neither surface runoff nor sediment load was measured on the small plots.

Tab.4 shows absolute and relative (%) runoff values. They can be compared with results of rainfall simulation experiments (GOCZAN et al. 1982). the experiments made use of the KAZP RAINFALL SIMULATOR (KAZO 1966) on an eroded sol brun lessivé under maize (see tab.5).

There is a big difference between the results of the plot and rain simulation

Plot	23.6.1982	23.7.1982	28.5.1984		2.6.1984		9.6.1984		7.-11.8.1984		9.1984		10.1984	20.5.1985	
	r	r	r	s	r	s	r	s	r	s	r	s	r	r	s
L	9.1	4.5	2.8	650	3.3	1050	6.8	2180	7.5		9.7		4.5	11.3	311
L_1				166		9		7800							597
L_2				208		56		7100		497		47			1232
L_3				283		78		1760		194		46			48
1															804
2			0.6	8.2	0.4	6									194
3			0.3	31	0.8	6									
4															274
5															530

r = runoff (l)
s = sediment (g)
L_1 = upper sieve, L_2 = middle sieve, L_3 = lower sieve
L = large plot
1–5 = small plots (see also fig.2)
the amount of sediments (s) given in the line of L means sediment deposited down in the sludge-tank

Tab. 3: *Runoff and sediment measurements 1982–1985.*

	23.6.1982	23.7.1982	28.5.1982	2.6.1982	9.6.1984	7.–11.8.1984	9.1984	10.1984	20.5.1985
precipitation (mm)	9.5	19	15.5	17	21.5	36.6	76.4	51.1	88.6
intensity (mm)	6	3	5	5.5	9	-	-	-	-
runoff volume (l)	9.1	4.5	2.8	3.3	6.8	7.5	9.7	4.5	11.3
runoff coefficient (%)	0.14	0.03	0.03	0.03	0.05	0.03	0.02	0.01	0.02

Tab. 4: *Runoff measurements 1982–1985.*

measurements. This can be explained as follows:

1. Although the slope on which the plots were installed is also covered mainly by an eroded sol brun lessivé, the upper slope section is very strongly eroded and the lower section is overlaid by colluvium.

2. Soil moisture and the length of the plot were different in the two measurements.

3. the big difference is mainly due to the different crop type, i.e. to the soil protecting effect of the wheat.

Slope %	Rainfall intensity (mm/h)				
	5	10	20	30	40
	Runoff values in l				
5	0.1	0.3	1.2	2.6	4.5
12	0.1	0.9	6.4	11.1	14.0
17	0.3	2.2	10.6	17.1	22.2
25	1.3	5.4	17.6	22.8	32.7
30	1.3	5.6	17.6	24.6	35.5

Tab. 5: *Results of rainfall simulation experiments (after GÓZCÁN et al. 1982).*

Time	Ca	Mg	P	Na	K	Fe	Mn	S
8.11.1984	33.62	7.12	0.93	9.84	1.53	1.71	0.07	4.36
9.1984	22.68	1.82	0.93	6.99	2.46	1.78	0.92	1.59
20.5.1985	34.25	1.19	-	-	2.4	2.27	0.22	0.63

Tab. 6: *Concentrations of selected ions in surface-runoff (mg/l)*.

Time	NO_2 ppm	NO_3 ppm	NH_4 ppm	N (sum total) ppm	total sum of N in the measured quantity of runoff mg
28.5.1984	1.0	1.7	0.01	0.696	1.949
2.6.1984	1.0	5.4	0.24	1.711	5.646
9.6.1984	1.5	0.1	0.03	0.502	3.414
11.8.1984	0.24	3.26	0.20	0.962	7.215
9.1984	0.12	0.88	1.60	0.856	8.303
20.5.1985	0.12	0.88	1.60	1.477	16.690

Tab. 7: *N content of surface runoff samples (ppm)*.

The runoff measurements show that low intensity rains also cause considerable runoff (the highest measured intensity was 9 mm/hour, see tab.4). The influence of different crop-type on soil erosion could only be shown when comparing results of rainfall simulation experiments with those of plot measurements.

The long control plot (plot No. 1) failed to give expected results since the error due to infiltration along the boardplanks was high. The plot had an elongated shape; its sides parallel to slope gradient were rather long relative to its area. Infiltration into the gently sloping sections along the slope also increased.

5.2 Sediments

When comparing the two measurements taken very close to each other (2.6.84 and 9.6.84) it is surprising that there is a much bigger difference in the volume of sediment load than in the amount of surface runoff (see tab.4). A moister soil and an increased rainfall intensity explain the difference.

No general trend can be observed when the sediments trapped on the sieves are compared. However, it could be observed that the sediment deposited on the 2nd sieve (i.e. 2–0.25 mm) formed in one case the biggest and another time the second biggest fraction.

Very little sediment was recorded from the small plots in 1984. There is only one measurement from 1985, referring to considerable deposition of sediments along the slope.

Runoff and sediment samples were analysed in the laboratory (see tab.6 and 7). Not all of the samples could be examined due to lack of sufficient quantity or to proper storage. Due to the lack of top soil on the slope, the proportion of Ca^{++} in the runoff samples is relatively high, pointing to the erosion of loess from the middle and upper slope sections.

The chemical analysis of runoff samples suggests to the transport of fertilizers. In 1984 twice as much fertilizers were applied as in 1985. This can be seen in the results of the laboratory measurements (tab.7).

6 Conclusions

1. 1. Although only 4 measurements are available, the results seem to be characteristic for the hilly regions of Hungary. For the extrapolation of the results control measurements are needed.

2. The measurements prove that water, soil and fertilizer loss due to low intensity rainfall is also considerable, so that on the basis of the above discussed measurements soil conservation measures are necessary.

3. Soil conservation should include a crop rotation. The biggest problem is to avoid or to minimize surface runoff therefore a soil amelioration can be suggested including loosening of sub-soil and a replacement of organic material in top soil by swampsoil.

References

ERÖDI, B., HORVÁTH, V., KAMARÁS, M., KISS, A. & SZEKRÉNYI, B. (1965): Talajvédö gazdálkodás hegy - és dombvidéken. (Soil conservation in mountainous and hilly areas.) Mezögazdasági Kiadó, Budapest. 463 p.

GÓZÁN, L., SCHÖNER, I. & TARNAI, P. (1973): Uj tipusu berendezés a geomorfodinamikai folyamatok analiziséhez, talaj - és környezetvédelmi kontrolljához. (New measuring equipment for the analysis of geomorphodynamic processes and for controlling environmental protection). Földrajzi Értesitö 22, 479–482.

GÓZÁN, L., KERTÉSZ, A., LÓCZY, D., MOLNÁR, K. & TÓZSA, I. (1982): A pilismaróti öblözet mezögazdasági területének talajtani és talajvizgazdálkodási viszonyai. (Soils and groundwater conditions in the area around Pilismarót.) MTA FKI, budapest, 8 p. + 10 maps.

KERTÉSZ, A. (1983): Bodenerosion in Ungarn. Das Meßprogramm des Geographischen Forschungsinstitits der Ungarischen Akademie der Wissenschaften. (Soil erosion in Hungary. the measurement program of the Geographical Research Institut of the Hung. Acad. of Sci.) Mitteilungen der Deutschen bodenkundlichen Gesellschaft **38**, 649–656.

SCHMIDT, R.-G. (1979): Probleme der Erfassung und Quantifizierung von Ausmaß und Prozessen der aktuellen Bodenerosion / Abspülung auf Ackerflächen. Physiogeographica, **1**, 240 p. Basel.

STEFANOVITS, P. (1964): Talajpusztulás Magyarországon. Magyarázatok Magyarország eróziós térképéhez. (Soil erosion in Hungary. Explanations to the soil erosion map of Hungary.) OMMI, budapest, 58 p.

SZILÁRD, J. (1982): A lejtöfejlödés müszeres vizsgálatának néhány eredménye. (Some results of instrumental investigation of slope development.) Földrajzi Ertesitö **31**, 191–20..

Address of authors:
László Góczán and Adám Kertész
Geographical Research Institute of the Hungary
Academy of Sciences
Népköztársaság-utja 62
H-1062 Budapest

GEOMORPHIC FACTORS IN LOCATING SITES FOR TOXIC WASTE DISPOSAL

H. Lavee, Ramat-Gan

Summary

Concerning the toxic waste from the Israeli chemical industries, a decision has been made by the responsible authorities that controlled shallow burial is the preferable disposal method; and, that the disposal site should be in the northern Negev desert.

The paper presents seven geomorphic principles in locating toxic waste burial sites. The intent of these principles is to minimize any toxic waste pollution of the groundwater for the worst scenario of conditions. The geomorphic considerations take into account climatic conditions, topography, soils and rocks properties, surface runoff and hydrogeology.

Based on the geomorphic criteria, three environmentally safe burial sites were recommended.

Resumen

A propósito de los desperdicios tóxicos procedentes de las industrias químicas israelíes, las autoridades responsables creen que le mejor método de deshacerse de ellos es mediante un soterramiento poco profundo, y que el mejor lugar para ello es al norte del desierto del Negev. El artículo presenta siete principios geomorfológicos para localizar puntos en los que sepultar desperdicios tóxicos, con el propósito de minimizar cualquier posible contaminación del agua subterránea. Las consideraciones geomorfológicas tienen en consideración las condiciones climáticas, la topografía, las propriedades de rocas y suelos, la escorrentía superficial y la hidrogeología. En base a estos criterios se recomiendan tres lugares para un soterramiento que no implique peligros mediambientales.

1 Background and Aim of the Present Study

Toxic waste from the Israeli chemical industries is transported to a single national centre for treatment and disposal. This centre, at Ramat-Hovav, is 12 km south of Be'er-Sheva, (fig.1). With the majority of this toxic waste treated in evaporation ponds, the residue from these ponds, as well as the remaining hazardous wastes require safe disposal.

The decision made by governmental authorities, is that shallow burial of double-sided barrels containing the toxic waste is the preferable disposal method; and, that the disposal site should be in the northern Negev, within 80 km of

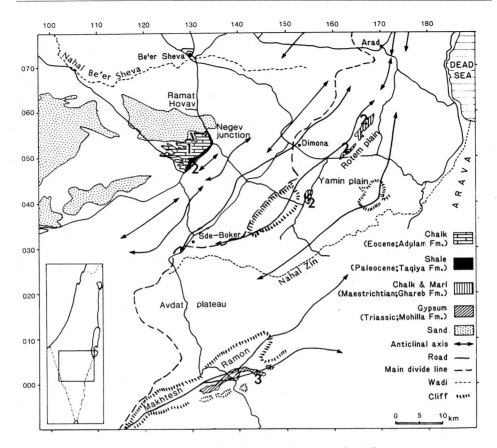

Fig. 1: *The preferred areas for waste burial.*

Ramat-Hovav. The final decision affecting the actual location of this site must include accessibility, tenure status, demographic, economic factors and, the environmental suitability of the site. This latter factor is strongly affected by geomorphic conditions.

2 Geomorphic Considerations

The choice of the northern Negev for the burial of toxic wastes was determined by the facts that a major portion of the main Israeli chemical industry is located in the vicinity, it is relatively low populated, and its physical characteristics make it one of the safest areas in the country.

The site's semi-arid climate with its strong seasonality, influences various parameters of the hydrological cycle, as follows:

1. Rainfall infiltrates usually to a depth of only a few meters; it eventually evaporates back into the atmosphere never reaching the water table.

2. Hillside overland flow is characterized by its discontinuity and only

small discharges reach adjoining channels (YAIR et al. 1980).

3. The main streams are ephemeral. Only occasionally, do intense storms cause sufficient overland flow to reach the channel to generate streamflow.

4. The regional aquifer is recharged from areas beyond the desert boundary and/or by the occasional flows in the main streams.

5. The regional water table is located several hundred meters below the surface.

Because of these hydrologic attributes, along with needs to protect populated areas, seven principles in locating toxic waste burial sites were formulated. The intent of these principles is to minimize any toxic waste pollution of the regional groundwater for the worst scenario of conditions.

1. Maximize the distance between the site and any main river channels, to prevent transportation of toxic wastes in surface flows or their infiltration into the groundwater.

2. Areas near catchment divides are preferable sites. The greater depth of the water table, the minimal likelihood of overland flow, and the low probability for overland flow reaching nearby channels is maximized in this topographic setting.

3. Areas with minimum gradients are preferred. The probability of overland flow reaching channels is low, as minimal wash occurs in these settings. In addition, these areas are more accessible to transportation.

4. Sites in which surface material is loose and characterized by high infiltration rates are recommended. This reduces the likelihood of surface flooding. This is on condition that the waste is buried deeply enough to prevent infiltrating water from reaching the waste. This unconsolidated material is advantageous also with regard to excavating and refilling the burial site.

5. The protection of the groundwater is enhanced if an aquitard protects the aquifer. It is best to bury the waste under the impervious material so that the vertical movement of the infiltrating water will be deflected to lateral flow before reaching the waste.

6. The site has to be in an area where the water table is at a great depth.

7. The site has to be distant from the densely settled coastal fringe of the Mediterranean and inland urban centres.

3 Method of the Study

Based on HATHEWAY & BLISS' recommendations (1982), the present study comprises a number of stages. First, using the criteria just presented, the geomorphic conditions of the northern Negev were characterized. The procedure was based on the analysis of maps and other relevant materials. Structure and lithology were ascertained from geological maps, scale 1:50,000 or 1:100,00. Aerial photographs and topographical maps were utilized to determine absolute and relative heights, relief intensity, drainage network, and drainage density. Geological cross-sections, information on major aquifers and local piezometric levels were taken from the literature.

Areas having impervious rock outcrops or impervious layers between the surface and the water table were delimited on maps. Those areas that also are topographically high and distant from the main channels were ranked as the preferred locations for burial sites.

4 Geomorphic Conditions of the Northern Negev

The northern Negev lies between Nahal Be'er Sheva Valley (200–300 m A.S.L.) in the north and Nahal Zin valley in the south. The eastern and central parts consist of a number of parallel anticlinal folds. These folds (500–600 m A.S.L.), largely conformable to the topography, are asymmetric: toward the northwest they are gentle (up to 12 degrees); toward the southeast they are steep to very steep (50 degree – 90 degree). The western part is built of low hills (350–450 m A.S.L.) and a few sandy areas. South of the Zin valley is the Avdat plateau and the Makhtesh Ramon (fig.1). The impervious layers in the northern Negev are comprised of gypsum, chalk, marl, and shales (fig.2).

5 Hydrology

5.1 Surface Runoff

The major Israel divide separates the northern Negev into two parts. The eastern part drains to the northern Arava and the Dead Sea and the western part drains to the Mediterranean. The western sandy areas are characterized by inefficient drainage (fig.1).

5.2 Groundwater

The major aquifers are:

1) The aquifer in rock formations of the Triassic age which is close to the surface at Makhtesh Ramon consists of limestone, dolomite and sandstone. Direct replenishment (pollution) from surface runoff is possible through the outcrops of these rocks.

2) The aquifer in rocks of the Cenomanian-Turonian age consists of limestone, dolomite and marl layers. This is a very important aquifer in central and northern Israel but only its southern fringes reach the northern Negev. The supply of fresh water to the southern part of the aquifer is from outcrops in the Hebron mountains with no replenishment from the Negev area.

3) The aquifer in rocks of the Eocene age is a perched aquifer where the formations of the Senonian-Paleocene age act as an aquiclude.

6 Preferred Sites for Toxic Waste Burial

Based primarily on geomorphic criteria, three areas were recommended:

1. Areas in which chalk layers of the Eocene age are exposed or are covered by several meters of sand, where marly chalk or shale formations (Menuha,Ghareb, Taqiya) of the Senonian-Paleocene age lies below (fig.3). Of these areas the hilly one which is near the Negev Junctions is preferred as it is close to Ramat-Hovav but still distant from the urban center of Be'er-Sheva (area no. 1 in fig.1).

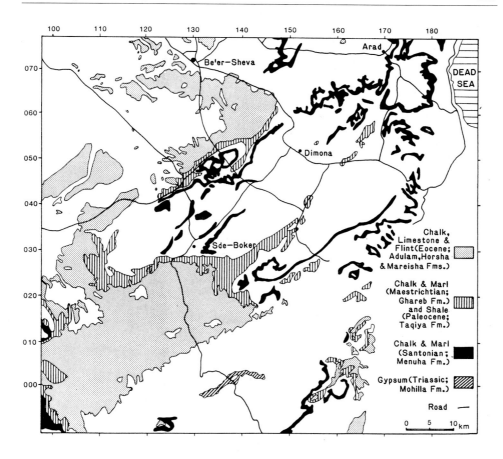

Fig. 2: *The impervious rock outcrops in the study area.*

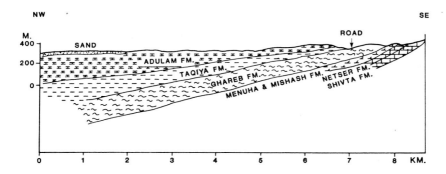

Fig. 3: *Geological cross-section west of the Negev Junction.*

2. Areas of the impervious rock outcrops of the Senonian-Paleocene age. Usually these are long and narrow outcrops parallel to the main structure lines. In many cases these outcrops are found in the floodplains of the major valleys and in such cases are not recommended. The Taqiya and Ghareb formations' outcrops near the Negev Junction and in the Yamin and Rotem plains are preferred (areas no. 2 in fig.1). In the Yamin and Rotem plains a conglomerate of several tens of meters of the Neogene age usually covers the Ghareb formation.

3. The hilly areas in the southern part of Makhtesh Ramon which consists of gypsum layers of the Mohilla formation (area no. 3 in fig.1).

Of the recommended areas and taking into account non-geomorphic considerations, authorities responsible for waste disposal must decide on the most attractive site. The selected site will be subjected to further analysis as to the following points:

1. Detailed mapping of the topography, faults, and the three-dimensional extent of the surficial geologic formations.

2. Field and laboratory testing of soil and rock samples in order to determine the amount of clay minerals, clay and silt size particles, relative density, porosity, and horizontal and vertical permeability.

3. Placement observation wells in order to define groundwater levels, quality and flow directions.

References

HATHEWAY, A.W. & BLISS, Z.F. (1982): Geomorphology as an aid to hazardous waste facility siting, northeast United States. In: R.G. Craig & J.L. Craft (eds.), Applied geomorphology. Allen & Unwin, London, 55–71.

U.S. ENVIRONMENTAL PROTECTION AGENCY (1975): Hazardous wastes: a review of literature and known approaches. 530/SW-161.

YAIR, A., SHARON, D. & LAVEE, H. (1980): Trends in runoff and erosion processes over an arid limestone hillside, northern Negev, Israel. Hydrol. Sci. Bull., **25(3)**, 243–255.

Address of author:
Hanoch Lavee
Department of Geography
Bar-Ilan University
Ramat-Gan, Israel

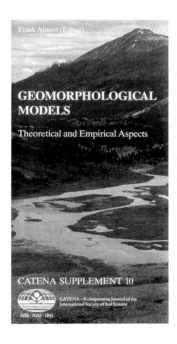

Frank Ahnert (Editor):

GEOMORPHOLOGICAL MODELS

Theoretical and Empirical Aspects

CATENA SUPPLEMENT 10, 1987

Price DM 149, — / US $88. —

ISSN 0722-0723 / ISBN 3-923381-10-7

CONTENTS

Preface

I. SLOPE PROCESSES AND SLOPE FORM

KIRKBY, M.J.
Modelling some influences of soil erosion, landslides and valley gradient on drainage density and hollow development.

TORRI, D.
A theoretical study of SOIL DETACHABILITY.

AI, N. & MIAO, T.
A model of progressive slope failure under the effect of the neotectonic stress field.

AHNERT, F.
Process-response models of denudation at different spatial scales.

SCHMIDT, K.-H.
Factors influencing structural landform dynamics on the Colorado Plateau – about the necessity of calibrating theoretical models by empirical data.

DE PLOEY, J. & POESEN, J.
Some reflections on modelling hillslope processes.

II. CHANNELS AND CHANNEL PROCESSES

SCHICK, A.P., HASSAN, M.A. & LEKACH, J.
A vertical exchange model for coarse bedload movement-numerical considerations.

ERGENZINGER, P.
Chaos and order – the channel geometry of gravel bed braided rivers.

BAND, L.E.
Lateral Migration of stream channels.

WIECZOREK, U.
A mathematical model for the geometry of meander bends.

III. SEDIMENT YIELD

YAIR, A. & ENZEL, Y.
The relationship between annual rainfall and sediment yield in arid and semi-arid areas. The case of the northern Negev.

ICHIM, I. & RADOANE, M.
A multivariate statistical analysis of sediment yield and prediction in Romania.

RAWAT, J.S.
Modelling of water and sediment budget: concepts and strategies.

MILLER, TH.K.
Some preliminary latent variable models of stream sediment and discharge characteristics.

IV. GENERAL CONSIDERATIONS

HARDISTY, J.
The transport response function and relaxation time in geomorphic modelling.

HAIGH, M.J.
The holon – hierarchy theory and landscape research.

TROFIMOV, A.M.
On the problem of geomorphological prediction.

SCHEIDEGGER, A.E.
The fundamental principles of landscape evolution.

J. Drescher, R. Horn & M. de Boodt (Editors):

Impact of Water and External Forces on SOIL STRUCTURE

CATENA SUPPLEMENT 11, 1988

DM 149,—/US $88.—

ISSN 0722-0723 / ISBN 3-923381-11-5

CONTENTS

Preface

W.F. Van Impe, M. De Boodt & I. Meyus
Improving the Bearing Capacity of Top Soil Layers by Means of a Polymer Mixture Grout

H.H. Becher
Soil Erosion and Soil Structure

H.-G. Frede, B. Chen, K. Juraschek & C. Stoeck
Simulation of Gas Diffusion

H. Bohne & R. Lessing
Stability of Clay Aggregates as a Function of Water Regimes

A.R. Dexter
Strength of Soil Aggregates and of Aggregate Beds

R. Horn
Compressibility of Arable Land

K.H. Hartge
The Reference Base for Compaction State of Soils

B.G. Richards & E.L. Greacen
An Example of Numerical Modelling – Expansion of a Root Cavity in Soil

A. Ellies
Mechanical Consolidation in Volcanic Ash Soils

H.H. Becher & W. Martin
Selected Physical Properties of Three Soil Types as Affected by Land Use

I. Håkansson
A Method for Characterizing the State of Compactness of an Arable Soil

C. Sommer
Soil Compaction and Water Uptake of Plants

W. Köppel
Dynamic Impact on Soil Structure due to Traffic of Off-Road Vehicles

W.E. Larson, S.C. Gupta & J.L.B. Culley
Changes in Bulk Density and Pore Water Pressure during Soil Compression

A.L.M. van Wijk & J. Buitendijk
A Method to Predict Workability of Arable Soils and its Influence on Crop Yield

N. Burger, M. Lebert & R. Horn
Prediction of the Compressibility of Arable Land

H. Borchert
Effect of Wheeling with Heavy Machinery on Soil Physical Properties

P.H. Groenevelt
Impact of External Forces on Soil Structure

B.P. Warkentin
Summary of the Workshop

R. B. Bryan (Editor)

RILL EROSION
Processes and Significance

CATENA SUPPLEMENT 8
192 pages / hardcover / price DM 149,— / US $ 88.—
Special rate for subscriptions until
December 15, 1987: DM 119,— / US $ 70.40

Date of publication: July 15, 1987   ORDER NO. 499/00107
ISSN 0722-0723/ISBN 3-923381-07-7

CONTENTS

R.B. BRYAN
PROCESSES AND SIGNIFICANCE OF RILL DEVELOPMENT

G. GOVERS
**SPATIAL AND TEMPORAL VARIABILITY IN RILL DEVELOPMENT
PROCESSES AT THE HULDENBERG EXPERIMENTAL SITE**

J. POESEN
TRANSPORT OF ROCK FRAGMENTS BY RILL FLOW—A FIELD STUDY

O. PLANCHON, E. FRITSCH & C. VALENTIN
RILL DEVELOPMENT IN A WET SAVANNAH ENVIRONMENT

R.J. LOCH & E.C. THOMAS
**RESISTANCE TO RILL EROSION: OBSERVATIONS ON THE
EFFICIENCY OF RILL EROSION ON A TILLED CLAY SOIL UNDER
SIMULATED RAIN AND RUN-ON WATER**

M.A. FULLEN & A.H. REED
**RILL EROSION ON ARABLE LOAMY SANDS
IN THE WEST MIDLANDS OF ENGLAND**

D. TORRI, M. SFALANGA & G. CHISCI
THRESHOLD CONDITIONS FOR INCIPIENT RILLING

G. RAUWS
**THE INITIATION OF RILLS ON PLANE BEDS OF
NON-COHESIVE SEDIMENTS**

D.C. FORD & J. LUNDBERG
**A REVIEW OF DISSOLUTIONAL RILLS IN LIMESTONE
AND OTHER SOLUBLE ROCKS**

J. GERITS, A.C. IMESON, J.M. VERSTRATEN & R.B. BRYAN
RILL DEVELOPMENT AND BADLAND REGOLITH PROPERTIES

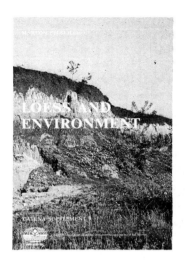

M. Pécsi (Editor)

LOESS AND ENVIRONMENT

SPECIAL ISSUE ON THE OCCASION OF THE XII th International Congress of the INTERNATIONAL UNION OF QUATERNARY RESEARCH (INQUA) Ottawa 1987

CATENA SUPPLEMENT 9
160 pages / hardcover / price DM 128,— / US $ 75.—
Special rate for subscriptions until
December 15, 1987: DM 102,40 / US $ 60.—

Date of publication: July 15, 1987   ORDER NO. 499/00 08
ISSN 0722-0723/ISBN 3-923381-08-5

CONTENTS

WANG Yongyan, LIN Zaiguan, LEI Xiangyi & WANG Shujie
FABRIC AND OTHER PHYSICO-MECHANICAL PROPERTIES OF LOESS IN SHAANXI PROVINCE, CHINA

J.P. LAUTRIDOU, M. MASSON & R. VOIMENT
LOESS ET GEOTECHNIQUE: L'EXEMPLE DES LIMONS DE NORMANDIE

A.J. LUTENEGGER
IN SITU SHEAR STRENGTH OF FRIABLE LOESS

WEN Qizhong, DIAO Guiyi & YU Suhua
GEOCHEMICAL ENVIRONMENT OF LOESS IN CHINA

W. TILLMANNS & K. BRUNNACKER
THE LITHOLOGY AND ORIGIN OF LOESS IN WESTERN CENTRAL EUROPE

A. VELICHKO & T.D. MOROZOVA
THE ROLE OF LOESS-PALEOSOLS FORMATION IN THE STUDY OF THE REGULARITIES OF PEDOGENESIS

H. MARUSZCZAK
STRATIGRAPHY OF EUROPEAN LOESSES OF THE SAALIAN AGE: WAS THE INTER-SAALIAN A WARM INTERSTADIAL OR A COLD INTERGLACIAL?

J. BURACZYNSKI & J. BUTRYM
THERMOLUMINESCENCE STRATIGRAPHY OF THE LOESS IN THE SOUTHERN RHINEGRABEN

M. PÉCSI & Gy. HAHN **PALEOSOL STRATOTYPES IN THE UPPER PLEISTOCENE LOESS AT BASAHARC, HUNGARY**

A.G. WINTLE
THERMOLUMINESCENCE DATING OF LOESS

A. BILLARD, E. DERBYSHIRE, J. SHAW & T. ROLPH
NEW DATA ON THE SEDIMENTOLOGY AND MAGNETOSTRATIGRAPHY OF THE LOESSIC SILTS AT SAINT VALLIER, DROME, FRANCE

G. COUDE-GAUSSEN & S. BALESCU
ETUDE COMPAREE DE LOESS PERIGLACIAIRES ET PERIDESERTIQUES: PREMIERS RESULTATS D'UN EXAMEN DES GRAINS DE QUARTZ AU MICROSCOPE ELECTRONIQUE A BALAYAGE

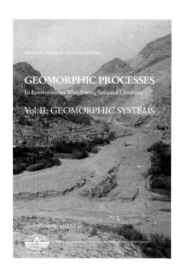

Adrian M. Harvey & Maria Sala:

GEOMORPHIC PROCESSES

In Environments With Strong
Seasonal Contrasts
Vol. II: GEOMORPHIC SYSTEMS

CATENA SUPPLEMENT 13, 1988

Price: DM 126,— / US $74.—

ISSN 0722-0723 / ISBN 3-923381-13-1

CONTENTS

Preface

M.A. Romero-Díaz, F. López-Bermúdez, J.B. Thornes, C.F. Francis & G.C. Fisher
Variability of Overland Flow Erosion Rates in a Semi-arid Mediterranean Environment under Matorral Cover, Murcia, Spain — 1

R.B. Bryan, I.A. Campbell & R.A. Sutherland
Fluvial Geomorphic Processes in Semi-arid Ephemeral Catchments in Kenya and Canada — 13

N. Clotet-Perarnau, F. Gallart & C. Balasch
Medium-term Erosion Rates in a Small Scarcely Vegetated Catchment in the Pyrenees — 37

M. Gutiérrez, G. Benito & J. Rodríguez
Piping in Badland Areas of the Middle Ebro Basin, Spain — 49

H. Suwa & S. Okuda
Seasonal Variation of Erosional Processes in the Kamikamihori Valley of Mt. Yakedake, Northern Japan Alps — 61

F. Gallart & N. Clotet-Perarnau
Some Aspects of the Geomorphic Processes Triggered by an Extreme Rainfall Event: The November 1982 Flood in the Eastern Pyrenees — 79

P. Ergenzinger
Regional Erosion: Rates and Scale Problems in the Buonamico Basin, Calabria — 97

M. Sorriso-Valvo
Landslide-related Fans in Calabria — 109

A.M. Harvey
Controls of Alluvial Fan Development: The Alluvial Fans of the Sierra de Carrascoy, Murcia, Spain — 123

C. Sancho, M. Gutiérrez, J.L. Peña & F. Burillo
A Quantitative Approach to Scarp Retreat Starting from Triangular Slope Facets, Central Ebro Basin, Spain — 139

A.J. Conacher
The Geomorphic Significance of Process Measurements in an Ancient Landscape — 147

NEW

CATENA paperback

Joerg Richter

THE SOIL AS A REACTOR
Modelling Processes in the Soil

If we are to solve the pressing economic and ecological problems in agriculture, horticulture and forestry, and also with "waste" land and industrial emmissions, we must understand the processes that are going on in the soil. Ideally, we should be able to treat these processes quantitatively, using the same methods the civil engineer needs to get the optimum yield out of his plant. However, it seems very questionable, whether we would use our soils properly by trying to obtain the highest profit through maximum yield. It is vital to remember that soils are vulnerable or even destructible although or even because our western industrialized agriculture produces much more food on a smaller area than some ten years ago.

This book is primarily oriented on methodology. Starting with the phenomena of the different components of the soils, it describes their physical parameter functions and the mathematical models for transport and transformation processes in the soil. To treat the processes operationally, simple simulation models for practical applications are included in each chapter.

After dealing in the principal sections of each chapter with heat conduction and the soil regimes of material components like gases, water and ions, simple models of the behaviour of nutrients, herbicides and heavy metals in the soil are presented. These show how modelling may help to solve problems of environmental protection. In the concluding chapter, the problem of modelling salt transport in heterogeneous soils is discussed.

The book is intended for all scientists and students who are interested in applied soil science, especially in using soils effectively and carefully for growing plants: applied pedologists, land reclamation and improvement specialists, ecologists and environmentalists, agriculturalists, horticulturists, foresters, biologists (especially microbiologists), landscape planers and all kinds of geoscientists.

Prof.Dr. Joerg Richter
Institute of Soil Science
University of Hannover, FRG

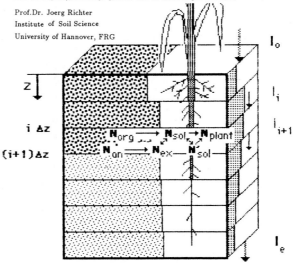

ISBN 3-923381-09-3   Price: DM 38,50 / US $ 24.—

CATENA

AN INTERDISCIPLINARY JOURNAL OF
SOIL SCIENCE
HYDROLOGY - GEOMORPHOLOGY
FOCUSING ON
GEOECOLOGY AND LANDSCAPE EVOLUTION

founded by H. Rohdenburg

A Cooperating Journal of the International Society of Soil Science (ISSS).

CATENA publishes original contributions in the fields of

GEOECOLOGY,
the geoscientific-hydro-climatological subset of process-oriented studies of the present ecosystem,

– the total environment of landscapes and sites

– the flux of energy and matter (water, solutes, suspended matter, bed load) with special regard to space-time variability

– the changes in the present ecosystem, including the earth's surface,
and

LANDSCAPE EVOLUTION,
the genesis of the present
ecosystem, in particular the genesis of its structure concerning soils, sediment, relief, their spatial organization and analysis in terms of paleo-processes;

– soils: surface, relief and fossil soils, their spatial organization pertaining to relief development,

– sediment with relevance to landscape evolution, the paleohydrologic environment with respect to surface runoff, competence, and capacity for transport of bed material and suspended matter, infiltration, groundwater and channel flow,

– the earth's surface, relief elements and their spatial – hierarchical organization in relation to soils and sediment

– the paleoclimatological properties of the sequence of paleoenvironments

---

ORDER FORM: Please, send your orders to your usual supplier or to:

USA/CANADA: CATENA VERLAG
                P. O. BOX 368
                Lawrence, KS 66044
                USA
                phone (913) 843-12 34

Other countries: CATENA VERLAG
                  Brockenblick 8
                  D-3302 Cremlingen
                  West Germany
                  phone 0 53 06/15 30
                  fax 0 53 06/15 60

CATENA 1988: Volume 15 (6 issues)

☐ please, enter a subscription 1988
at US $ 235.— / DM 398.—
incl. postage & handling

☐ please, send a free sample copy of CATENA

☐ please, send guide for authors

☐ please, enter a personal subscription 1988 at 50 % reduction
(available from the publisher only)

☐ I enclose | check | bank draft | unesco coupons |

☐ charge my credit card (only for orders USA/CANADA)
☐ Master Card   ☐ Visa

Card No. _____
Expir. Date _____
Signature _____

☐ please, send invoice

Name _____
Address _____
Date/Signature _____

NEW SOIL TECHNOLOGY

A Cooperating Journal of **CATENA**

SOIL TECHNOLOGY

This quarterly journal is concerned with applied research and field applications on

- soil physics,
- soil mechanics,
- soil erosion and conservation,
- soil pollution,
- soil restoration.

The majority of the articles will be published in English but original contributions in French, German or Spanish, with extended summaries in English will occasionally be considered according to the basic principles of the publisher CATENA whose name not only represents the link between different disciplines of soil science but also symbolizes the connection between scientists and technologists of different nations, different thoughts and different languages.

The coordinator of SOIL TECHNOLOGY:

Donald Gabriels,
Faculty of Agricultural Sciences, State University of Gent,
Coupure links 653,
B-9000 Gent, Belgium (tel 32-91-236961).

Editorial Advisory Board:

J. Bouma, Wageningen, The Netherlands
W. Burke, Dublin, Ireland
S. El. Swaify, Hawaii, USA
K. H. Hartge, Hannover, F.R.G.
M. Kutilek, Praha, CSSR
G. Monnier, Montfavet, France
R. Morgan, Silsoe, UK
D. Nielsen, Davis, Californ., USA
I. Pla Sentis, Maracay, Venezuela
J. Rubio, Valencia, Spain
E. Skidmore, Manhattan, Kansas, USA

**Editorial Office
SOIL TECHNOLOGY**

Dr. D. Gabriels
Department of Soil Physics
Faculty of Agriculture
State University Gent
Coupure Links 653
B-9000 Gent
Belgium
tel. 32-91-236961

**Papers published in
Vol. 1, No. 1, March 1988**

S. A. El Swaify, A. Lo, R. Jay, L. Shinshiro, R. S. Yost: Achieving conservation-effectiveness in the tropics using legume intercrops.

I. Pla Sentis: Riego y desarrollo de suelos afectados por sales en condiciones tropicales. / Irrigation and development of salt affected soils under tropical conditions.

K. H. Hartge: Erfassung des Verdichtungszustandes eines Bodens und seiner Veränderung mit der Zeit. / Techniques to evaluate the compaction of a soil and to follow its changes with time.

M. Kutilek, M. Krejča, R. Haverkamp, L. P. Rendon, J. Y. Parlange: On extrapolation of algebraic infiltration equations.

M. Šir, M. Kutilek, V. Kuráž, M. Krejča, F. Kubik: Field estimation of the soil hydraulic characteristics.

J. Albaladejo Montoro, R. Ortiz Silla, M. Martinez-Mena Garcia: Evaluation and mapping of erosion risks; an example from S. E. Spain.

SHORT COMMUNICATIONS

D. Gabriels: Use of organic waste materials for soil structurization and crop production; initial field experiment.

K. Reichardt: Aspects of soil physics in Brazil.

P. Bielek et al.: Internal nitrogen cycle processes and plant responses to the band application of nitrogen fertilizers.

V. Chour: An actual demand for improved soil technology in irrigation and drainage design in Czechoslovakia.

BOOK REVIEWS

ORDER FORM:

Please, send your orders to your usual supplier or to:

USA/CANADA: CATENA VERLAG
P.O.BOX 368
Lawrence, KS 66044
USA
phone (913) 843-12 34

Other countries: CATENA VERLAG
Brockenblick 8
D-3302 Cremlingen
West Germany
phone 0 53 06/15 30
fax 0 53 06/15 60

SOIL TECHNOLOGY   1988: Volume 1 (4 issues)

- ☐ please, enter a subscription 1988 at US $ 120.— / DM 198,— incl. postage and handling
- ☐ please, send a free sample copy of **SOIL TECHNOLOGY**
- ☐ please, send guide for authors
- ☐ please, enter a personal subscription 1988 at 50 % reduction (available from the publisher only)
- ☐ I enclose | check | bank draft | unesco coupons |
- ☐ charge my credit card (only for orders USA/CANADA)
  - ☐ Master Card   ☐ Visa

Card No. _____
Expir. Date _____
Signature _____
☐ please, send invoice
Name _____
Address _____
Date/Signature _____